Biogeochemistry
of a Forested Ecosystem

Gene E. Likens   F. Herbert Bormann

Biogeochemistry of a Forested Ecosystem

Second Edition

With 49 Illustrations

Springer-Verlag
New York Berlin Heidelberg London Paris
Tokyo Hong Kong Barcelona Budapest

Gene E. Likens
Institute of Ecosystem Studies
Millbrook, NY 12545-0129
USA

F. Herbert Bormann
School of Forestry and
Environmental Studies
Yale University
New Haven, CT 06511
USA

Cover photograph courtesy of Judith Meyer, University of Georgia.

Library of Congress Cataloging-in-Publication Data
Likens, Gene E., 1935–
Biogeochemistry of a forested ecosystem/Gene E. Likens, F. Herbert Bormann.—2nd ed.
p. cm.
Rev. ed. of: Biogeochemistry of a forested ecosystem/Gene E. Likens . . . [et al.]. c1977, 160 pp.
Includes bibliographical references and index.
ISBN 0-387-94502-4.—ISBN 3-540-94502-4 (hardcover)
ISBN 0-387-94351-X.—ISBN 3-540-94351-X (softcover)
1. Forest ecology—New Hampshire—Hubbard Brook Experimental Forest. 2. Biogeochemistry—New Hampshire—Hubbard Brook Experimental Forest. 3. Hubbard Brook Experimental Forest (N.H.) I. Bormann, F. Herbert, 1922– . II. Title. III. Title: Biogeochemistry of a forested ecosystem.
QH105.NL554 1995
574.5′2642′097423—dc20 94-41866

Printed on acid-free paper.

Acquiring editor: Robert C. Garber.
Production coordinated by Chernow Editorial Services, Inc., and managed by Francine McNeill; manufacturing supervised by Jacqui Ashri.
Typeset by Best-set Typesetter Ltd., Hong Kong.
Printed and bound by Edwards Brothers, Inc., Ann Arbor, MI.
Printed in the United States of America.

9 8 7 6 5 4 3 2 1

ISBN 0-387-94502-4 Springer-Verlag New York Berlin Heidelberg (hardcover)
ISBN 0-387-94351-X Springer-Verlag New York Berlin Heidelberg (softcover)
ISBN 3-540-94351-X Springer-Verlag Berlin Heidelberg New York (softcover)

Preface to the Second Edition

When we originally published *Biogeochemistry of a Forested Ecosystem* in 1977, the Hubbard Brook Ecosystem Study (HBES) had been in existence for 14 years, and we included data through 1974, or a biogeochemical record of 11 years. Now our continuous, long-term biogeochemical records cover more than 31 years, and there have been many changes. The most notable change, however, is that three of our coauthors on the original volume are now deceased. They are deeply missed in so many ways.

In spite of the longer records, different trends, and new insights, we believe that the basic concepts and approaches we presented in 1977 represent the most valuable contribution of the original edition. They are still valid and useful, particularly for an introductory study of, or course in, biogeochemistry. Our goal in this revision is to preserve these features, correct errors, and revise or eliminate misleading or ambiguous short-term data (11 years!), while maintaining approximately the original length and the modest cost.

We have left the data of the original text largely unchanged; a few figures and tables have been replaced, revised, or added. We provide selected references to guide the interested reader to some of the longer-term data and newer research of the HBES. To accomplish these objectives, each chapter has a brief addendum of especially interesting new approaches, important long-term trends, and new insights gained from the longer-term perspective (31 years). We specifically draw attention to a few of the HBES articles published since 1977 for a more detailed description and analysis. A complete listing of the publications from the HBES can be found in "Publications of the Hubbard Brook Ecosystem Study" (Likens, 1994). Data from the HBES are available on a publicly accessible system (computer modem telephone number 603-868-1006) maintained by the U.S.D.A. Forest Service and the HBES.

As in the original version of this book, we do not include or consider here data from the experimentally manipulated (e.g., by clearcutting) watershed-ecosystems at Hubbard Brook, which is covered in our book, *Patterns and Processes in a Forested Ecosystem* (Bormann and Likens, 1979).

We thank Phyllis C. Likens for help in preparing the manuscript for this revision. We also acknowledge the help of Donald C. Buso with data analysis. Major financial support for the HBES since 1977 has been provided by the National Science Foundation and the Andrew W. Mellon Foundation.

September 1994

Gene E. Likens
F. Herbert Bormann

Preface to the First Edition

About 15 years ago we began the Hubbard Brook Ecosystem Study with the development of an ecosystem model and the conception of a method whereby some major parameters of the model could be directly measured under field conditions. The method, called "the small watershed technique," allowed measurement of input and output of chemicals and the construction of ecosystem nutrient budgets. Although the Hubbard Brook study of nutrient cycling originated with ideas developed by F.H. Bormann and G.E. Likens, its early growth was aided by Robert Pierce, forest hydrologist; Noye Johnson, geochemist; and John Eaton, forest ecologist. Donald W. Fisher of the U.S. Geological Survey also cooperated in the early phases of the project and provided numerous data on the chemistry of precipitation and stream water. Particular credit is due the U.S. Forest Service, whose scientists chose the Hubbard Brook Valley as a hydrologic study site, selected particular watersheds for intensive measurement, carried out a variety of basic hydrologic studies, and in general cooperated with us in many ways to make the Hubbard Brook Ecosystem Study a reality.

The initial part of the ecosystem study was concerned primarily with nutrient flux and cycling, and it was done slowly and deliberately. The entire effort during the first few years of study was carried forward by three of us at Dartmouth College, with the cooperation of the U.S.D.A. Forest Service. We had no precedents to follow, because similar, comprehensive studies of natural ecosystems had not been done. We reasoned that it would be best to first construct a solid base of studies on nutrient–hydrologic interactions upon which subsequent studies could be built. In this regard, we were fortunate to rather quickly determine quantitative nutrient budgets for replicated ecosystems. These results gave us guiding insight into the function of natural ecosystems and into the development of future lines of research.

Nevertheless, we felt that slower growth would be more manageable and would allow for substantial interaction among all senior investigators to ensure proper coordination and development of the overall study.

Generally, the method we used to guide the growth of the Hubbard Brook research proposals was as follows: First, based on our own perceptions and feedback from ongoing studies, from cooperating scientists, and from outside advisors, we identified research problems that were timely and particularly pertinent to our overall goals. Some of the studies were launched under our direction; others were brought to the attention of an established investigator working in that area. We then sought a mutually satisfactory arrangement to allow the investigator to work at Hubbard Brook. From the beginning, we have called attention to the kinds of information a cooperative study might produce, but we have always encouraged individuality in the design and execution of research. *We deem this individual research freedom one of the greatest assets of the Hubbard Brook Study*. Individuality in selection of problems and conduct of research is also encouraged among graduate students at Hubbard Brook. Not only does this contribute to the intellectual ferment and sound growth of the Hubbard Brook Study, but as educators, we feel this approach absolutely necessary if the Hubbard Brook Study is to contribute to graduate education.

Over the last 15 years the study has grown to include not only hydrology and input–output chemistry, but also aspects of the structure, function, and dynamics of the forest ecosystem itself. In addition to studies of undisturbed ecosystems, entire ecosystems have been experimentally manipulated to allow comparative study of undisturbed and perturbed ecosystems. Stream and lake ecosystems within the Hubbard Brook Valley have been studied in detail, and the connection between the forest ecosystem and these interlinked aquatic ecosystems is now the focus of considerable attention.

Approximately 50 senior scientists and scores of graduate students have carried out studies at Hubbard Brook during the last 15 years. Over 200 publications have resulted from this work.

We now plan to organize some of this information in ways we consider useful both to the scientists concerned with the theory of biogeochemical cycles and the structure, function, and development of forested ecosystems, and to the land-use specialists who are concerned with the production of a variety of benefits, goods, and services from northern hardwood ecosystems.

We have developed two volumes so far, and a third is planned for the future. This, the first volume, presents a detailed examination of the biogeochemistry of an undisturbed, aggrading, second-growth northern hardwood forest at Hubbard Brook. Ecosystems similar to this cover a large area of northern New England and New York. Major emphasis in this book is on the physical aspects of nutrient and hydrologic flow through the ecosystem and nutrient budgetary. Not only are we concerned in this book with the presentation of data and conclusions that may be drawn from them, but we also discuss methodology as it influences results, and we try to share some of our experience at Hubbard Brook in the hope that other workers can avoid pitfalls where we have not been so fortunate. Moreover, we hope that the data presented in this volume can serve as a basis for decision-making and management schemes, including modeling within the northern hardwood forest landscape.

The second volume has as its primary concern the structure, function, and development through time of the northern hardwood ecosystem and is designed for the advanced student in ecology as well as for the ecosystem specialist. It concentrates on the interrelationships among biogeochemical processes, structure, and species behavior within the ecosystem and how these change with time following perturbation. We plan for the information in these two volumes to interrelate and thus draw support for each other. Biogeochemistry is presented more fully in this first volume, whereas biological considerations are dealt with more fully in the second volume.

By no means do these two volumes summarize the Hubbard Brook Study. In the future we hope to put together a more comprehensive volume that involves contributions from the many cooperating scientists who have brought their particular specialties to bear on the Hubbard Brook Ecosystem.

There are many persons to acknowledge and thank for contributions to such a multidisciplinary study conducted over so many years. Many data were collected during rainstorms, blizzards, intense cold, darkness, and numerous other types of northern New Hampshire weather. Long, often tedious hours in the laboratory and at the computer center were also required. The ideas, devotion, loyalty, and encouragement and genuine enthusiasm of field technicians, laboratory technicians, students, and various other colleagues made the overall Hubbard Brook Ecosystem Study a reality and is beyond a simple "thank you."

For this little book, we owe debts great and small to many people—to those whose ideas and encouragement provided an intellectual thrust to move ahead; to those without whose devotion, skills, and endurance there would have been no progress at all; and to those mostly good-natured colleagues whose shafts of humor and criticisms have helped us all to keep a realistic perspective. For all of these things and more, we thank:

J. Aber, B. Ackerman, M. Alexander, T. Allison, H. Art, P. Art, S. Bennett, R. Bilby, J. Bishee, C. Black, A. Bock, H. Bond, B. Bormann, R. Bormann, D. Botkin, E. Brastrup, J. Brink, S. Brown, A. Brush, L. Burckes, T. Burton, D. Buso, P. Cantino, J. Chamberlain, S. Chisholm, B. Coffin, C. Cogbill, B. Corson, W. Covington, M. Cushman, H. Davis, M. Davis, F. deNoyelles, G. deSylva, A. Dominski, J. Duggin, F. Edington, Y. Falusi, A. Federer, N. Fernando, S. Fiance, A. Field, D. Fisher, S. Fisher, L. Forcier, M. Fox, J. Fryer, G. Furnival, J. Galloway, A. Gambell, D. Gerhart, J. Gilbert, W. Glanz, J. Gosz, C. Grant, M.

Gross, R. Hame, G. Hart, B. Haywood, J. Hobbie, R. Holmes, S. Holmgren, J. Hornbeck, S. Howe, J. Janak, P. Johnson, M. Jordan, F. Juang, H. Karsten, M. Keller, W. Kent, L. Kivell, J. Koonce, S. Koppes, D. Korbobo, M. Koterba, B. Kuehn, R. Lavigne, T. Ledig, J. Lehman, R. Leonard, V. Levasseur, E. Loker, B. Mahall, J. Makarewicz, P. Marks, W. Martin, D. Masza, K. McConnochie, O. MacGregor, D. McNaught, A. McPhail, J. Melillo, R. Mendelsohn, R. Miller, J. Meyer, R. Moeller, M. Mollony, D. Mulcahy, R. Muller, J. Murphy, P. Murtaugh, D. Nelson, J. Nobel, R. Patrick, P. Pelton, B. Peterson, N. (Crane) Poole, G. Potter, R. Reifsnider, W. Reiners, R. Reynolds, L. Riggs, J. Roskoski, D. Ryan, M. Schmale, P. Sargent, E. Sayetta, V. Scarlet, W. Schlesinger, J. Schorske, L. Schwartz, M. Sebring, T. Sherry, T. Siccama, D. Smith, W. Smith, P. Sprague, D. Sprugel, D. Stout, S. Strom, F. Sturges, J. Tinker, P. (Fruend) Tinker, J. Thompson, G. Thornton, D. Tiernan, P. Toyryla, K. Turekian, G. Voigt, J. Wallis, B. Walsh, R. Walter, P. Wicklund, R. Withington, G. Whitney, R. Whittaker, T. Wood, G. Woodwell, J. York, and C. Zoltov.

We all acknowledge continued indulgence from our and our coauthors' wives (Kay, Chris, Carol, Diane, and Delores) and families for time spent away from them.

Financial support for the field studies was provided by the National Science Foundation and the United States Forest Service. This work was done through the cooperation of the Northeastern Forest Experiment Station, Forest Service, United States Department of Agriculture, Upper Darby, Pennsylvania.

The manuscript was initially formulated while Gene E. Likens was a John Simon Guggenheim Fellow at the Australia National University. This support is gratefully acknowledged.

January 1977

Gene E. Likens
F. Herbert Bormann

Contents

1
Ecosystem Analysis

An ecological system has a richly detailed budget of inputs and outputs of energy and matter. Because of the lack of precise information about these relationships and the internal functions that maintain the ecosystem, it is often difficult to assess the impact of human activities on the biosphere. As a result, land-use planners often cannot take into account or even foresee the full range of consequences a project may have. Without full information, the traditional practice in the management of land resources has been to emphasize strategies that maximize the output of some desirable product or service and give little or no thought to the secondary effects. As a result one sees such ecological confusion as an all-out effort to increase food production while natural food chains become increasingly contaminated with pesticides and runoff and seepage waters carry increasing burdens of pollutants from fertilizers and farm wastes. Forests are cut with inadequate perception of the effects on regional water supplies, wildlife, recreation, and esthetic values, and wetlands are converted to commercial use with little concern over important hydrologic, biologic, esthetic, and commercial values lost in the conversion.

For some time it has been evident that a new conceptual approach to the management of resources is desirable. One approach that has received considerable attention recently is to consider entire ecological systems as single interacting units, instead of limiting studies to only a few components of the system. In an experimental forest in New Hampshire we have been conducting a large-scale investigation aimed at supplying this kind of information that is usually lacking about ecosystems.

A vast number of variables, including biologic structure and diversity, geologic heterogeneity, climate, and season, control the flux of both water and chemicals through ecosystems. Clearly, both living and nonliving components of ecosystems are important in defining and regulating the flow of matter within and between ecosystems. Because chemicals tend to circulate from nonliving components to living organisms and back to the nonliving environment, the pathways have been termed

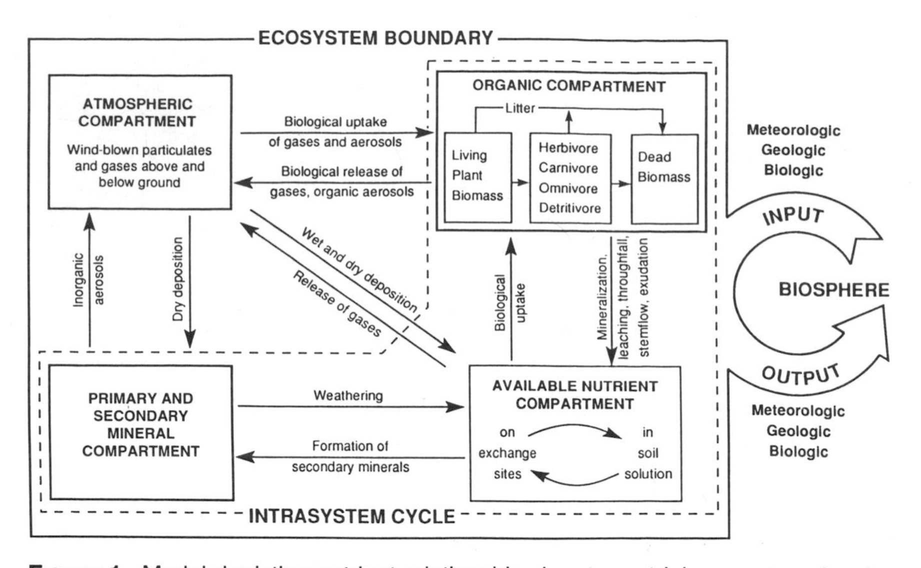

FIGURE 1. Model depicting nutrient relationships in a terrestrial ecosystem. Inputs and outputs to the ecosystem are moved by meteorologic, geologic and biologic vectors (Bormann and Likens, 1967; Likens and Bormann, 1972). Major sites of accumulation and major exchange pathways within the ecosystem are shown. Nutrients that, because they have no prominent gaseous phase, continually cycle within the boundaries of the ecosystem between the available nutrient, organic matter and primary and secondary mineral components tend to form an intrasystem cycle. Fluxes across the boundaries of an ecosystem link individual ecosystems with the remainder of the biosphere. Modified from Likens, 1992.

biogeochemical cycles. Ecosystems continually exchange (gain or lose) matter and energy with other ecosystems and the biosphere as a whole and, therefore, biogeochemical cycles are vital to the maintenance of the system. Comparisons of biogeochemical data from natural ecosystems with those that have been manipulated by humans or otherwise disturbed then provide important information about the functional efficiency or "health" of an ecosystem.

For more than three decades the biogeochemistry of the Hubbard Brook Experimental Forest, New Hampshire, has been the subject of continuous study (Figure 1). More than 20,000 water samples, including rain, snow, stream water, ground water, throughfall, and stemflow, have been collected and chemically analyzed. These extensive data now provide the means and the perspective to identify, isolate, and quantify some of the complex and interlocking biogeochemical processes of a forested ecosystem. As this study shows, seasonal and year-to-year variations in flux rates and associated biologic responses often require long-term analysis before reliable generalizations can be drawn about such complex natural systems. This book summarizes our understanding of the biogeochemical flux of water and nutrients through an aggrading northern hardwood forest ecosystem, refers to the time scale at which the various biogeochemical processes operate, and relates these data to other ecosystems throughout the world.

We began studies in 1963 to determine the magnitude of the biogeochemical flux and internal cycling of nutrients in northern hardwood forest ecosystems in the White Mountains of New Hampshire. Our ecosystem unit is a watershed or drainage area, with vertical and horizontal boundaries defined functionally by biologic activity and the drainage of water (Bormann and Likens, 1967). Six adjacent, small watersheds within the Hubbard Brook Experimental Forest, hereafter referred to as HBEF, with similar vegetation and geology and subject to the same climate, have been used for these studies (Bormann and Likens, 1967; Likens and Bormann, 1972). This series of watershed ecosystems provided the possibility for replication as well as a design for experimental manipulation of entire natural ecosystems.

A conceptual model of these deciduous forest ecosystems was developed to facilitate the quantitative evaluation of the input–output flux and cycling of water and chemicals (Bormann and Likens, 1967; Likens and Bormann, 1972). The continuous flow of energy, water, nutrients, and other materials across the ecosystem's boundaries are considered to be inputs and outputs, which are transported by meteorologic, geologic, and biologic vectors (Figure 1). Meteorologic inputs and outputs consist of wind-borne particulate matter, dissolved substances in rain and snow, aerosols, and gases (e.g., CO_2). Geologic flux includes dissolved and particulate matter transported by surface and subsurface drainage, and the mass movement of colluvial materials. Biologic flux results when chemicals or energy gathered by animals in one ecosystem are deposited in another (e.g., local exchange of fecal matter or mass migrations). These input–output categories are therefore defined as vectors or "vehicles" for transport of nutrients, matter, or energy rather than sources; i.e., a leaf blown into an ecosystem represents meteorologic input rather than biologic input.

Within the ecosystem, the nutrients may be thought of as occurring in any one of four basic compartments: (1) atmosphere, (2) living and dead organic matter, (3) available nutrients and (4) primary and secondary minerals (soil and rock). The atmospheric compartment includes all elements in the form of gases or aerosols both above and below ground. Available nutrients are ions that are absorbed on or in the humus or clay–humus complex or dissolved in the soil solution. The organic compartment includes all nutrients incorporated in living and dead biomass. The woody tissues of living vegetation are considered a part of the living biomass. The primary and secondary minerals contain nutrients that comprise the inorganic soil and rock portions of the ecosystem.

The biogeochemical cycling of elements involves an exchange between the various compartments within the ecosystem. Available nutrients and gaseous nutrients may be taken up and assimilated by the vegetation and microorganisms; some may be passed on to heterotrophic consumers and then made available again through respiration, biologic decomposition, and/or leaching from living and dead organic matter. Insoluble primary and secondary minerals may be converted to soluble available nutrients through the general process of weathering; soluble nutrients may be redeposited as secondary minerals.

Small Watershed Approach

In humid regions chemical flux and cycling are intimately linked to the hydrologic cycle. Hence one cannot measure the input and output of nutrients without simultaneously measuring the input and output of water. The problem usually is that subsurface flows of water, which can be a significant fraction of the hydrologic cycle, are almost impossible to measure.

About 31 years ago it occurred to us that under certain circumstances the interaction of the nutrient cycle and the hydrologic cycle could be turned to good advantage in the quantitative study of an ecosystem. The requirements were that the ecosystem be a watershed underlain by tight bedrock or some other impermeable base. In that case the only inputs would be meteorologic and biologic; geologic input need not be considered because there is no transfer between adjacent watersheds. In humid areas where surface wind is a minor factor, losses from the system are only geologic and biologic. Given an impermeable base, all the geologic output would inevitably turn up in the streams draining the watershed. When the watershed is part of a larger and fairly homogeneous biotic unit, the biologic output tends to balance the biologic input because most animals move randomly into and out of the watershed, randomly acquiring or discharging nutrients. One therefore need measure only the meteorologic input and the geologic output of nutrients to arrive at the net gain or loss of a given nutrient in the ecosystem. Using this approach, it has been possible to obtain quantitative budgets for most of the major chemical nutrients. The flux of some nutrients with a gaseous phase (e.g., N and S) is more difficult to evaluate, but the ecosystem approach has enabled us to make realistic assumptions and estimates about some of these unmeasured gaseous components. We have used this approach at Hubbard Brook in an intensive study of six contiguous watersheds within the HBEF (Figure 2). These watershed ecosystems are all tributary to Hubbard Brook.

The Hubbard Brook Ecosystem

The HBEF was established in 1955 by the U.S.D.A. Forest Service as the principal research area for the management of watersheds in New England. The name of the area is derived from the major drainage stream in the valley, Hubbard Brook. Hubbard Brook flows generally from west to east for about 13 km (Figure 2) until it joins with the Pemigewasset River, which ultimately forms the Merrimack River and discharges into the Atlantic Ocean. Water from more than 20 tributaries enters Hubbard Brook along its course. Mirror Lake, a small oligotrophic lake, discharges into Hubbard Brook at the lower end of the valley.

Location

The HBEF is located near West Thornton within the White Mountain National Forest of north central New Hampshire. Coordinates of 43°56′ N, 71°45′ W bisect the area. The Atlantic Ocean is about 116 km to the southeast.

Climate

Although the climate varies with altitude, it is classified as humid continental with short, cool summers and long, cold winters (Trewartha, 1954). The climate may be characterized by (1) changeability of the weather, (2) a large range in both daily and annual temperatures, and (3) equable distribution of precipitation. HBEF lies in the heart of the middle latitudes and the majority of the air masses therefore flow from west to east. During the winter months these are northwesterlies and during the summer the air generally flows from the southwest. Therefore, the air affecting HBEF is predominantly continental. However, during the autumn and winter, as the colder polar air moves south, cyclonic disturbances periodically move up the east coast of the United States providing an occasional source of maritime air. The mean air temperature in July is 19°C and in January is −9°C (Federer, 1973). A continuous snowpack develops each winter to a depth of about 1.5 m. Occasionally, mild temperatures in midwinter partly or wholly melt the snowpack. A significant microclimatologic feature of this area is that even the uppermost layer of the forest soils usually remains unfrozen during the coldest months because of the thick humus layer and a deep snow cover (Hart et al., 1962).

Area, Topography, and Aspect

The HBEF covers an area of 3,160 ha and ranges in altitude from 222 to 1,015 m. The experimental watershed ecosystems (Watersheds 1–6, Figure 2) range in size from 12 to 43 ha and in altitude from 500 to 800 m. These

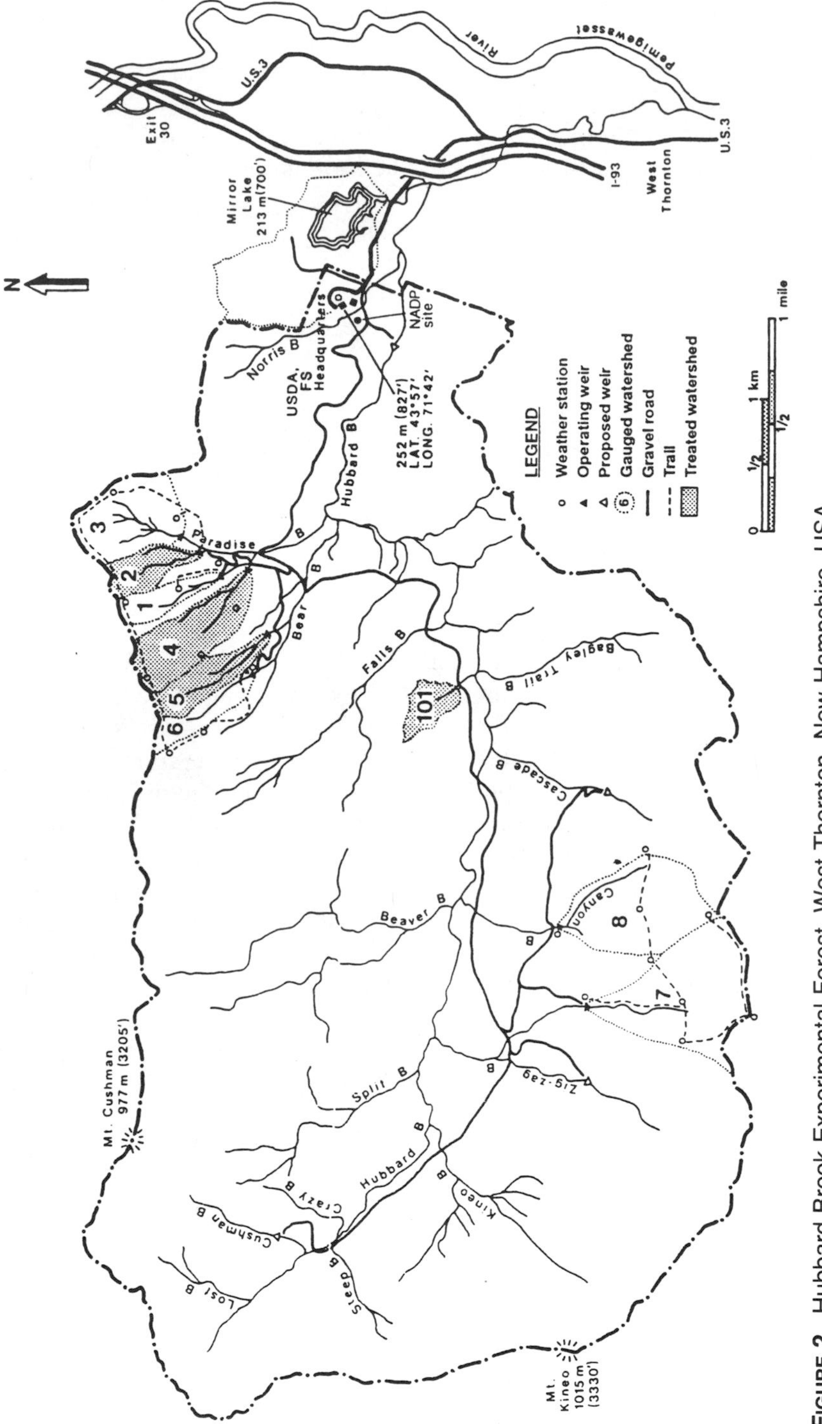

FIGURE 2. Hubbard Brook Experimental Forest, West Thornton, New Hampshire, USA.

headwater watersheds are all steep (average slope of 20–30%) and face south. The experimental watersheds have relatively distinct topographic divides. The height of the land surrounding each watershed ecosystem and the area have been determined from ground surveys and aerial photography.

Geology

The geologic substrate, outcrops of bedrock and stoney till, in the Hubbard Brook Valley was exposed some 12,000–13,000 years ago when the glacial ice sheet retreated northward (Likens and Davis, 1975). The eastern portion of the Hubbard Brook Valley is underlain by a complex assemblage of metasedimentary and igneous rocks. The major component is the Silurian Rangeley formation, consisting of quartz-mica schist and quartzite interbedded with sulfidic schist and calc-silicate rock. Originally deposited as pelites, sandstones, and conglomerates, these rocks were metamorphosed to sillimanite grade and have undergone four distinct phases of deformation. The metaphoric rocks also were intruded by a variety of igneous rocks, including the Concord Granite, Spaulding Quartz Diorite, pegmatite, and diabase and lamprophyre dikes. The western portion of the valley is underlain by the Devonian Kinsman pluton, a foliated, granitic rock distinquished by large phenocrysts of potassium feldspar. Much of the area of the experimental watersheds is covered with glacial till derived from local bedrock. Losses of surface water by deep seepage through the bedrock are considered minimal (Likens et al., 1967).

Soils

Soils are mostly well-drained spodosols (haplorthods) of sandy loam texture, with a thick (3–15 cm) organic layer at the surface. Most precipitation infiltrates into the soil at all times and there is very little overland flow (Pierce, 1967). This is because the soil is very porous, the surface topography is very rough (pit and mound, mostly from windthrown trees), and normally there is little soil frost.

Soil depths are highly variable but average about 0.5 m from surface to bedrock or till. Soil on the ridges may consist of a thin accumulation of organic matter resting directly on the bedrock. In some places, impermeable pan layers at depths of about 0.6 m restrict vertical water movement and root development. The soils are acid (pH $\leqslant$ 4.7) and generally infertile.

Vegetation and Fauna

The vegetation of the HBEF is part of the northern hardwood ecosystem, an extensive forest type that extends with variations from Nova Scotia to

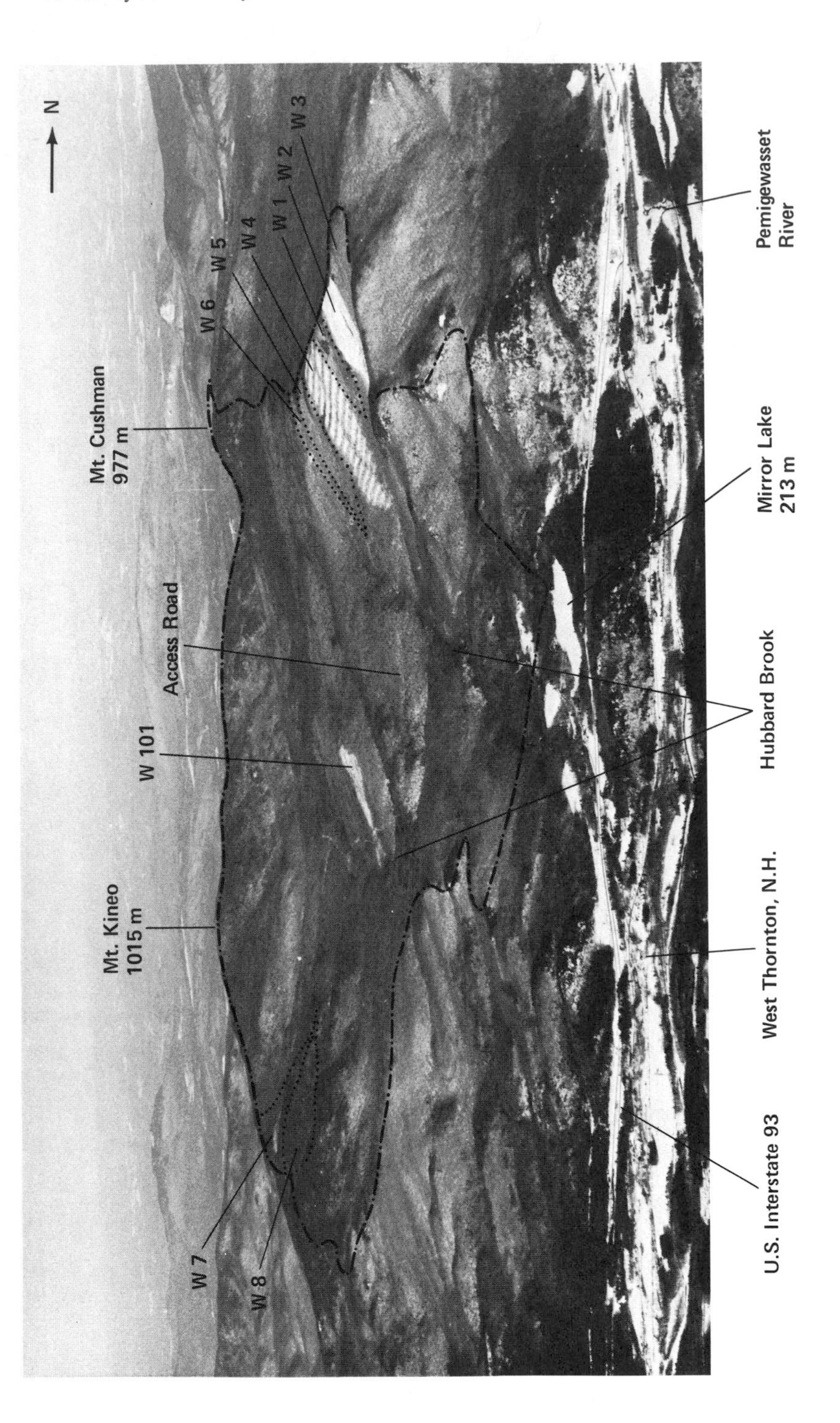

N
Mt. Cushman
977 m
W 6
W 5
W 4
W 1
W 2
W 3
Access Road
W 101
Mt. Kineo
1015 m
W 7
W 8
Pemigewasset
River
Mirror Lake
213 m
Hubbard Brook
West Thornton, N.H.
U.S. Interstate 93

the western Lake Superior region and southward along the Blue Ridge Mountains (Braun, 1950; Küchler, 1964; Oosting, 1956). Classification of mature forest stands as northern hardwood ecosystems rests on a loosely defined combination of deciduous and coniferous species that may occur as deciduous or mixed deciduous–evergreen stands. Principal deciduous species include, beech (*Fagus grandifolia*), sugar maple (*Acer saccharum*), yellow birch (*Betula alleghaniensis*), white ash (*Fraxinus americana*), basswood (*Tilia americana*), red maple (*Acer rubrum*), red oak (*Quercus borealis*), white elm (*Ulmus americana*); the principal coniferous species are hemlock (*Tsuga canadensis*), red spruce (*Picea rubens*), and white pine (*Pinus strobus*) (Braun, 1950). At Hubbard Brook, the vegetation is characteristic of a developing, northern hardwood forest ecosystem. The forest was logged in 1910–1919, damaged by a hurricane in 1938, but there is no evidence of fire in the history of these stands (Bormann et al., 1970; Bormann and Likens, 1979). The forest is unevenly aged and well stocked, primarily with sugar maple (*Acer saccharum*), American beech (*Fagus grandifolia*), and yellow birch (*Betula alleghaniensis*). Red spruce (*Picea rubens*), Balsam fir (*Abies balsamea*), white birch (*B. papyrifera* var. *cordifolia*), and hemlock (*Tsuga canadensis*) are prominent on north-facing slopes, on ridge tops, on rock outcrops, and along the main channel of Hubbard Brook. The forest had a basal area of about $24\,m^2$/ha in 1965 (Bormann et al., 1970). About 100 species of shrubs and herbs may be found in the watershed ecosystems (Siccama et al., 1970). Aboveground biomass accumulation for the forest averaged about 4.85 Mg/ha·yr during 1965–1977, but declined to 0.89 Mg/ha·yr during 1987–1992 (Likens et al., 1994).

Fauna common to northern hardwood forests occur within the Hubbard Brook Valley. More than 90 species of birds have been observed (Likens, 1973) and snowshoe hare (*Lepus americanus*), beaver (*Castor canadensis*), red fox (*Vulpes fulva*), black bear (*Ursus americanus*), moose (*Alces americana*), and whitetail deer (*Odocoileus virginianus*) are

◀———

PHOTOGRAPH 1. *Aerial view of Hubbard Brook.* Hubbard Brook Experimental Forest, West Thornton, New Hampshire (area within –·– symbol, approximately 3,076 ha). Experimental watersheds shown by dotted boundaries (see Figure 2). Watershed 2 was cleared of all woody vegetation in 1965 and sprayed with herbicides for three successive years to prevent regrowth. No forest products were removed. Watershed 4 was cut in alternating strips in three phases over a 4-yr period beginning in 1970. Picture shows Watershed 4 when two-thirds of the strips were cut and harvested. Watershed 101 was cleared and products were removed in one operation in 1970. Watershed 5 was cut (whole-tree harvest) in 1983–1984 after this photograph was taken. All other watersheds are untreated as yet and serve as references. Dark areas on the ridges and peaks generally are conifers and the lighter areas are deciduous northern hardwoods.

PHOTOGRAPH 2. *A spodosol soil developed on till at the HBEF.* The uppermost dark layer, an organic layer called the forest floor, rests on the mineral soil. Immediately below the forest floor is a bleached and chemically leached horizon, which grades into a zone of accumulated organic matter and iron and aluminum oxides. At still lower depths, the soil varies from partially weathered to unweathered mineral components. The presence of an obvious leached horizon is intermittent within the HBEF.

present. The White Mountains generally have a low population density of deer because of severe winters and high hunting pressure (Siegler, 1968).

Drainage Streams

Headwater streams draining the watershed ecosystems are small and perennial. Flow ranges from zero during infrequent summer droughts to hundreds of m^3/ha·day during snowmelt and storm events. Two of the watersheds, 3 and 4 (42 and 36 ha, respectively), are larger and typically maintain flow during summer (Figure 2).

From June to October the streams are heavily shaded by the forest vegetation. As a result daily water temperatures do not vary more than a degree or two Celsius. The annual temperature range in these streams is from 0 to about 18°C. The stream water is normally saturated or supersaturated with dissolved oxygen and the pH is less than 6.

The stream bed is covered by organic debris, fine sand, gravel, cobbles, and small boulders. There are occasional bedrock exposures in the stream channel and these are preferred sites for gaging weirs. Numerous dams of organic debris occur throughout the length of the stream; the stream therefore has a stair-step appearance of alternating "waterfalls" and pools. The stream channel comprises about 1–2% of the area of each watershed ecosystem.

Additional details concerning the topography, climate, geology, and biology of the HBEF are given elsewhere (Bormann et al., 1970; Holmes and Sturges, 1973; Likens et al., 1967; Johnson et al., 1968; Likens, 1973; Marks, 1974; Northeastern Forest Experiment Station, 1972; Siccama et al., 1970; Whittaker et al., 1974). The results of our earlier biogeochemical studies have been reported by Bormann and Likens (1967), Likens et al. (1967), Fisher et al. (1968), Johnson et al. (1968, 1969), Juang and Johnson (1967), Bormann et al. (1969, 1974), Pierce et al. (1970), Likens et al. (1971), Likens and Bormann (1972), Fisher and Likens (1973), Hobbie and Likens (1973), Marks (1974), Sturges et al. (1974), Hornbeck et al. (1970, 1975a,b), Burton and Likens (1975), and Likens and Bormann (1975). In addition to providing basic data on ecosystem function and the ecological relationships between structure and function, these biogeochemical data provide a baseline for judging the effects of human manipulations of forested landscapes (e.g., Likens et al., 1970; Pierce et al., 1972).

The next chapters, Hydrology through Nutrient Cycles, present detailed information about the long-term biogeochemistry of the HBEF. These sections summarize our current understanding of the flux and cycling of chemicals in the watershed ecosystems at Hubbard Brook. An attempt is made to synthesize results and to interpret and comment on outstanding problems, solved and unsolved.

Photograph 3. *Hubbard Brook at an elevation of about 215 m.* The characteristic boulder substrate and "stair-step" nature of this stream are evident.

Addendum

An Experimental Application of the Small Watershed Technique

The Hubbard Brook "sandbox experiment" (Bormann et al., 1987) was designed to apply the small watershed technique to outdoor ecosystem-mesocosms lined and filled with soil substrates of known volume: soil organic matter (SOM), and primary and secondary minerals. The purpose was to allow experimental exploration of ecosystem processes, like biological nitrogen fixation and weathering of primary minerals, that were particularly difficult to study in mass-balance techniques in complex naturally occurring watershed ecosystems.

For example, in the case of nitrogen (N), sandboxes were filled with homogenized sandy soil of known N content, planted with seeds or seedlings of known N content, and allowed to revegetate or were maintained free of vegetation (Bormann et al., 1993). After an appropriate time period, changes in N storage were measured as (Δ soil) and (Δ vegetation). Nitrogen in vegetation included litter, herb, and tree biomass, including roots, to the bottom of the sandbox. Additionally, input of N (N bp) was estimated from the Hubbard Brook bulk precipitation record; output of N was measured directly in drainage water from a representative sandbox (N dr).

Changes in storage were calculated with the equation:

$$\Delta \text{ N storage} = \Delta \text{ N soil} + \Delta \text{ N veg.} \tag{1}$$

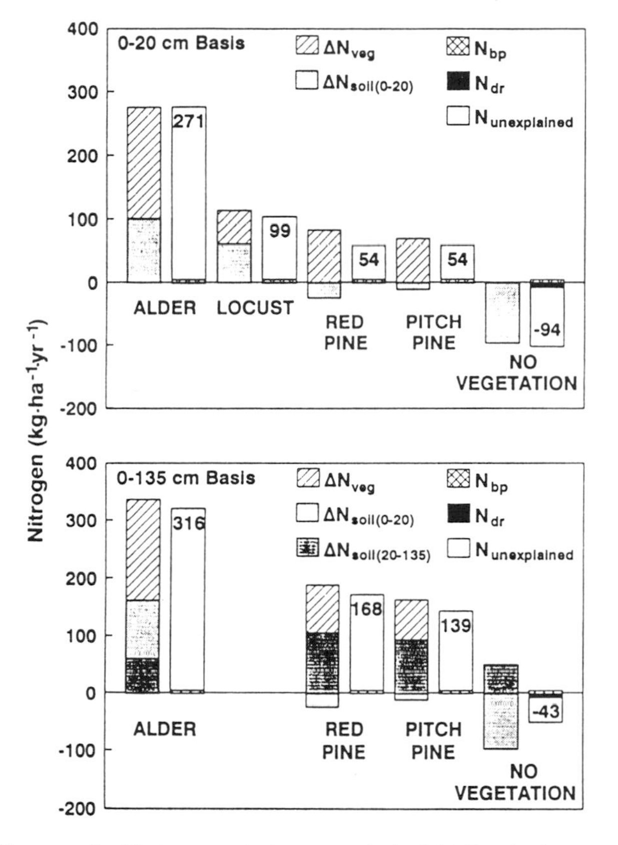

Figure 3. Changes in N storage during a period of 4–5 yr in the top 20 cm (top panel) and in the top 135 cm (bottom panel) in vegetation and soil (left bar for each species) and average annual inputs, outputs, and $N_{unexplained}$ (right bar). Numbers in the right bar are the values for $N_{unexplained}$. $N_{unexplained}$ for alder (*Alnus* spp.) and locust (*Robina* spp.) are attributed to symbiotic N fixation while those for red (*P. resinosa*) and pitch (*P. rigida*) pine are for associative N fixation. From Bormann et al., 1993.

To determine if changes in N storage were balanced by measured N inputs minus outputs, the following equation was used:

$$\Delta \text{ N storage} = \text{N bp} - \text{N dr}. \quad (2)$$

Rewriting Eq. 2 to estimate what fraction of N storage could not be explained by the mass balance equation, ($N_{unexplained}$), produced:

$$N_{unexplained} = \Delta \text{ N storage} - \text{N bp} + \text{N dr} \quad (3)$$

Probable sources of $N_{unexplained}$ included biological N_2 fixation (N bnf), dry deposition (N dd), and biological and chemical volatilization (N vol) as follows:

$$N_{unexplained} = \text{N bnf} + \text{N dd} - \text{N vol} \quad (4)$$

The experimental sandbox approach, combined with biogeochemical data, provides an accurate method for evaluating N budgets for experimental ecosystems under field conditions (Figure 3). It also provides opportunities for studying plant–soil interactions. In sandbox studies at Hubbard Brook, it was found that N added in precipitation and lost in drainage water was quantitatively unimportant relative to biological N fixation. Preventing vegetation from becoming established in a sandbox resulted in substantial losses of N in drainage water and, perhaps still greater losses by N volatilization. These studies also revealed a surprisingly high rate of associative nitrogen fixation in sandboxes dominated by *Pinus* suggesting the need for accelerated research on associative N_2 fixation in other pines and other conifers.

Currently the sandbox approach at Hubbard Brook is being used to evaluate the rate at which primary minerals are weathered and the rate at which nutrients are added to the available nutrient pool as an ecosystem undergoes temporal development.

2 Hydrology

Because of the vital role of water as a transporting agent, chemical solvent, and catalyst, quantitative data on hydrology are of paramount importance in understanding the biogeochemistry of an ecosystem. The U.S.D.A. Forest Service has monitored and maintained accurate records of precipitation and streamflow for watersheds of the HBEF since 1956.

The Water-Year

Selection of a suitable water-year is a primary consideration for hydrologic analyses. The ideal water-year is that successive 12-month period that most consistently, year after year, gives the highest correlation between precipitation and streamflow. In such watersheds as the HBEF, streamflow is largely dependent on (1) precipitation; (2) the amount of water storage opportunity in the soil and thereby the amount of water stored in the soil at any time; and (3) the amount of water stored in the snowpack.

After a large array of linear regressions for successive 12-month periods of precipitation and streamflow had been examined, the water-year beginning on 1 June and ending on 31 May was chosen. This water-year exhibited correlation coefficients of about 0.99. This high correlation can be explained by several phenomena: (1) Although evapotranspiration each summer leaves the soil in variable states of water content, autumn rains completely replenish water depleted during the summer. (2) Water storage opportunity for the watershed as a whole is on the order of about 10–15 cm; and (3) the addition of between 20 and 30 cm of water to the soil during melting of the snowpack is sufficient to fully recharge the soil. The selection of 1 June water-year is also advantageous because this corresponds to the period when foliage is getting established. This allows for a separation of the water-year into periods essentially coincident with growing and dormant seasons of the vegetation.

Precipitation

Because precipitation serves as a major vehicle of nutrient input into the ecosystem, its accurate measurement is of prime importance in an evaluation of any biogeochemical cycle. Precipitation is measured within the area of the experimental watersheds by a fixed network of standard United States Weather Bureau precipitation gages, approximately one for every 13 ha, in conjunction with continuous weighing and recording collectors. The Thiessen polygon method (Thiessen, 1911) is used to determine amounts of areal precipitation over entire watersheds.

On the average 129.5 cm of water falls on each unit area of watershed annually. Of this some 62%, or 80 cm, becomes streamflow and the remainder, about 49 cm, is lost as vapor through evapotranspiration (Table 1). The evapotranspiration is estimated as the difference between precipitation and streamflow for the water-year because the geologic base of the watershed is watertight (see the section on Deep Seepage, below).

There are on the average 111 precipitation days (>trace) per year at Hubbard Brook, or about two per week. Dividing the average annual precipitation of 129.5 cm by 111 days gives an average intensity of 1.2 cm per precipitation day.

The concept of a normal or average water-year is of limited utility. Although the long-term average was 129.5 ± 15.8 cm/yr during 18 yr of record (1956–1974), in only 9 yr has precipitation been within ±10 cm of this value (Figure 4). The mean deviation from the average precipitation is 12.2% at Hubbard Brook and is less than the expected mean variability, about 16%, for similar rainfall areas of the world (Conrad, 1941). In the driest water-year on record, 1964–1965, precipitation was 95.1 cm, whereas in the wettest year, 1973–1974, it was 185.7 cm. These values represented the extremes of wet and dry conditions relative to precipita-

TABLE 1. Average Annual Hydrologic Budget for the Hubbard Brook Experimental Forest.

	cm/unit area $\pm s_{\bar{x}}$	Percent of total
1956–1974		
Input (precipitation)	129.5 ± 5.1	100
Output (streamflow)	80.1 ± 5.0	61.9
Output (evaporation and transpiration)	49.4 ± 0.79	38.1
1963–1974 (period of biogeochemical study)		
Input (precipitation)	132.2 ± 6.8	100
Output (streamflow)	83.3 ± 6.6	63
Output (evaporation and transpiration)	48.9 ± 1.0	37

tion; they were 27% less and 43% greater, respectively, than the long-term average at Hubbard Brook. This range of 90.6 cm emphasized the need for a long-term record to understand the ecosystem's hydrologic flux. If the biogeochemistry of our ecosystems had been based upon either of these years alone, our estimates of meteorologic input and stream-water output of nutrients would have been substantially misleading. As it turned out, during the period of our biogeochemical studies, 1963–1974, annual precipitation averaged 132.2 ± 14.9 cm/yr and this period included both the wettest and driest years of record. Despite the occurrence of both "extreme" wet and dry years during the period of our study (1963–1974), the standard deviation of the mean precipitation was only 11% and the range varied by less than twofold. This relative "stability" of the precipitation for the HBEF imparts stability to the chemical budgets, as well, because the two are interdependent.

In any particular year monthly amounts of precipitation may show random extremes, but for the longer term, the monthly pattern becomes quite regular (Figures 5 and 6). That is, the average amount of precipitation is relatively constant on a monthly basis throughout the year. On the

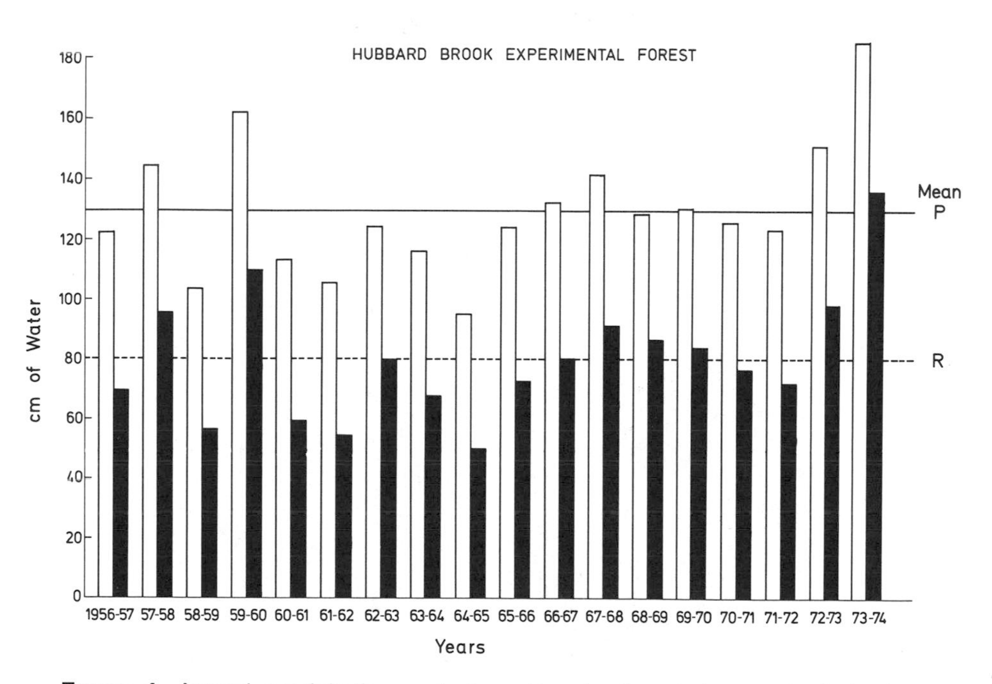

FIGURE 4. Annual precipitation and streamflow for forested watersheds in the Hubbard Brook Experimental Forest. The horizontal lines indicate the average values during 1956–1974. *P*, mean precipitation; *R*, mean streamflow; *ET*, mean evapotranspiration; $ET = P - R$, solid bars show annual streamflow; open bars show annual precipitation. Streamflow from manipulated or disturbed watersheds is not included in these results. Each water-year is from 1 June to 31 May.

Photograph 4a–d. *Precipitation collectors.* Various types of precipitation collectors used at the HBEF. Collectors are located in clearings within the forest to avoid contamination. No overhead obstruction is allowed above an angle of 45° from the opening of the collector. Collectors are positioned at least 2 m above the ground. (**a**) Gages for quantitatively monitoring amounts of precipitation (rain and snow). *Left*: standard United States Weather Bureau nonrecording gage with

c

d

20.3-cm orifice. *Right*: weight-activated recording gage. Alter-type windshields reduce wind velocity around gage opening to provide a sample equivalent to deposition on ground. A mixture of ethylene glycol and methanol are added to gages in winter to convert snow to liquid state. (**b**) A polyethylene apparatus used for collecting rain for chemical analysis consisting of a 28-cm diameter funnel, tubing, 2-liter reservoir, and vapor barrier. Samples are retrieved and a clean apparatus installed at 1-week intervals. (**c**) An automatic sampler for the collection and separation of wet and dry depositions for chemical analysis. A sensor on the lid activates the mechanism to move the lid from one bucket to the other at the beginning of wet or dry periods. (**d**) Plastic garbage can used to collect snow for chemical analysis. Samples are retrieved and a clean collector installed at 1-week intervals.

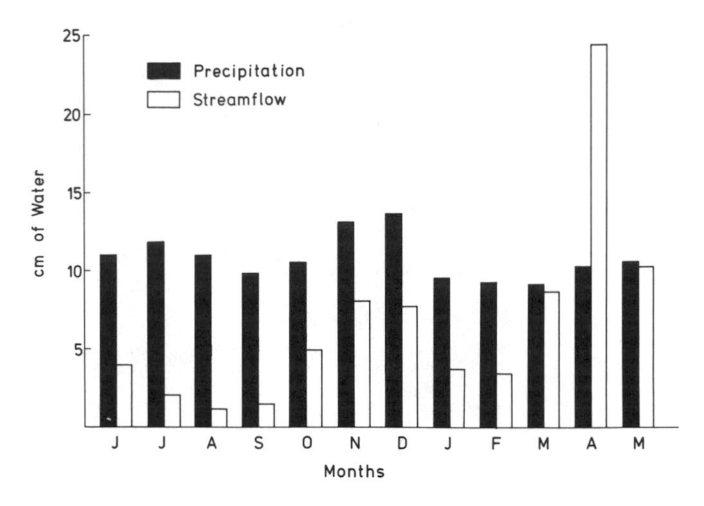

FIGURE 5. Average monthly precipitation and streamflow for the Hubbard Brook Experimental Forest during 1956–1974.

average, December is the wettest month and March is the driest, but on the average no month deviates widely from the overall monthly average of 10.8 cm.

The seasonal effects of temperature on precipitation (rain or snow), as well as the structure and function of the biomass (e.g., interception of raindrops by the forest canopy and litter), are important in lowering the erosive potential of precipitation. During the winter most of the precipitation falls as snow with a much smaller potential energy of impact. Indeed, some 30% of the annual precipitation falls as snow (Table 2). The pro-

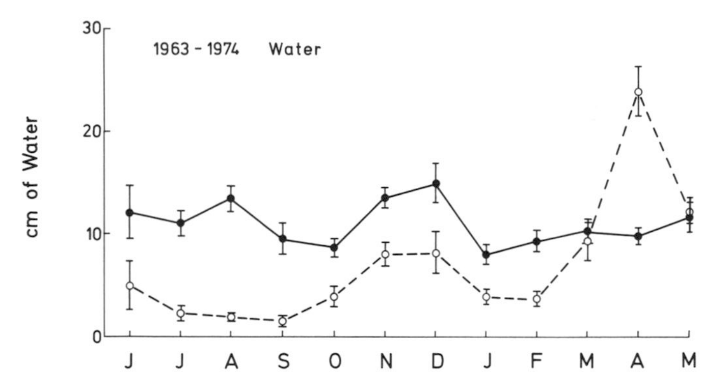

FIGURE 6. Average monthly precipitation (solid line) and streamflow (broken line) for the Hubbard Brook Experimental Forest during 1963–1974. The vertical bars are ± one standard deviation of the mean.

TABLE 2. Annual Input of Water as Snow or Rain for the Hubbard Brook Experimental Forest.

	Percent of annual time period	Percent of annual water input
Snow	41 (mid-Nov.–mid-April)	30 (38.9 cm)
Rain	59 (mid-April–mid-Nov.)	70 (90.6 cm)

portion of the year when snow is generally expected is larger than this (41%; Table 2) but some rain occurs in the winter months.

Streamflow and Evapotranspiration

Streamflow is measured continuously throughout the year by automatic recorders at stream-gaging stations, which include either a V-notch weir

PHOTOGRAPH 5. *Stream gaging weir.* Tandem arrangement of modified San Dimas flume and weir for monitoring streamflow on Watershed 6. The metal flume on the left measures high flows; the 90° sharp edged, V-notch weir on the right measures low flows. The gaging station is built on bedrock so that all streamflow is channeled first through the flume and then into the weir. Recording instruments in shelters for both flume and weir monitor streamflow continuously. Propane burners are used during the winter to prevent freezing. Large particulate matter transported by the stream during high flows is caught either in the screen as the flow cascades from the flume over the trough, bypassing the V-notch, or in the ponding basin behind the V-notch. Gaging stations similar to this or utilizing a V-notch weir only have been established on eight experimental watersheds in the HBEF (see Figure 2).

or a combination of V-notch and San Dimas flume anchored to the bedrock at the base of each watershed. Heating units in the ponding basin above the V-notch permit accurate measurements of streamflow in the coldest weather. In contrast to the monthly uniformity of precipitation, most of the streamflow at the HBEF occurs in spring when the accumulated snowpack melts (Figures 5 and 6). Some 54% of the annual streamflow occurs during March, April, and May, with more than 30% in April alone. During the summer months streamflow is very low, particularly during August and September (Figures 5 and 6). During this period water loss from the ecosystem occurs primarily through evapotranspiration. A secondary, minor peak in streamflow occurs in November and December when transpiration has virtually ceased after the loss of the deciduous leaves, and with somewhat greater autumn rainfall. At that time soil water deficits are eliminated and most of the rain moves through the soil and into the streams. Winter streamflow is diminished as precipitation accumulates as a snowpack.

On an annual basis, evapotranspiration remains relatively constant despite major fluctuations in precipitation. This is in contrast to annual streamflow, which is highly correlated with the amount of precipitation (Figure 7). This suggests that on an annual basis evapotranspiration is fairly constant, that it has first call on precipitation, and that once evapotranspiration is satisfied the remainder of the precipitation goes to streamflow. In a general way this appears to be true, but a more detailed analysis suggests a more complex situation. Streamflow during the months of June through September has an exponential relationship to precipitation, starting slowly and rising after about 45 cm (Figure 8). A simulation of evapotranspiration during the growing season, June through September (Figure 8), shows that evapotranspiration gradually increases to about 50 cm of precipitation and then levels off. Although these data are rough, because we do not know precisely the effect of stored soil water conditions at the end of May and at the end of September on streamflow and evapotranspiration during June through September, they suggest a certain reasonable logic about the hydrologic behavior.

The relationship between evapotranspiration and precipitation underscores the powerful regulating role that the living ecosystem plays in the hydrologic cycle. During the active growing season, evapotranspiration (mostly transpiration) is an important factor in regulating the amount of streamflow. Evapotranspiration then not only acts as a regulator for the hydrologic cycle but also contributes to tighter nutrient cycles by diminishing streamflow that carries nutrients out of the ecosystem.

Seasonal streamflow varies by orders of magnitude, whereas annual streamflow varies by no more than twofold and is usually less. Because the chemical flux is so intimately tied to the hydrologic cycle, the effects of seasonal hydrologic events have corresponding effects on the seasonal nutrient budgets. In contrast, the annual hydrologic budget, with its

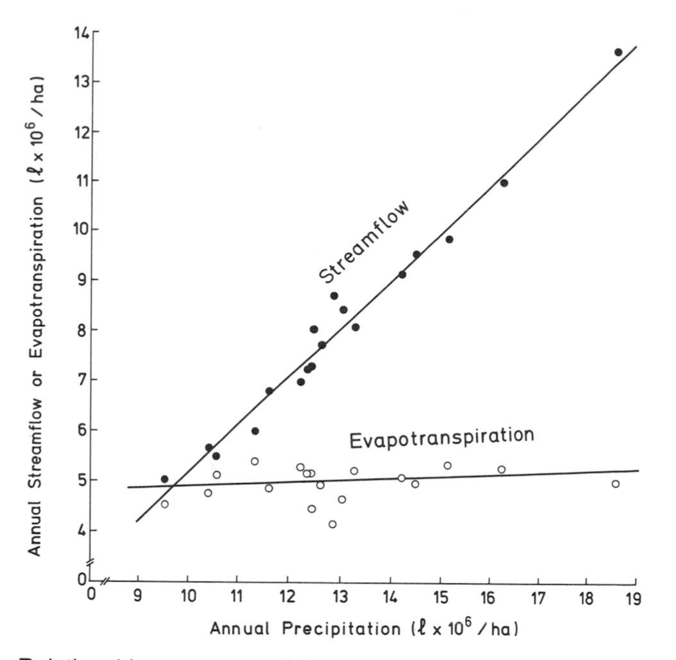

FIGURE 7. Relationship among precipitation, streamflow, and evapotranspiration for the Hubbard Brook Experimental Forest during 1956–1974. The regression lines fitted to these data are $Y = b + ax$, where Y = annual streamflow or evapotranspiration in 10^6 l/ha, b = Y intercept, a = slope, and x = annual precipitation in 10^6 l/ha. The equations for each line are: streamflow, $Y = 0.96x - 4.45$ (F ratio of 0.0001***; correlation coefficient, 0.99); evapotranspiration, $Y = 0.038x + 4.49$ (F ratio of 0.33; correlation coefficient, 0.24).

relatively small variability, imparts long-term stability to the associated annual chemical budgets, as is shown in subsequent sections.

A factor of prime importance in regulating erosion and stream-water chemistry at the HBEF is that the soils have a high infiltration capacity. Because the soil is very porous, virtually all incident precipitation, rain or snowmelt, percolates through the soil even in the largest storms. Surface runoff in the conventional sense is not manifested at HBEF except in a few boggy areas and in channels (Pierce, 1967). The downhill motion of water toward stream beds is by subsurface interflow. The soil usually remains unfrozen during the coldest months because of the thick humus layer and a deep snow cover (Hart et al., 1962). The surface topography is rough (pit and mound), which also facilitates infiltration rather than overland flow. Because the geologic substrate is essentially watertight and losses of water by deep seepage are minimal, almost all water has intimate contact with the biologic and inorganic components of the soil as it moves

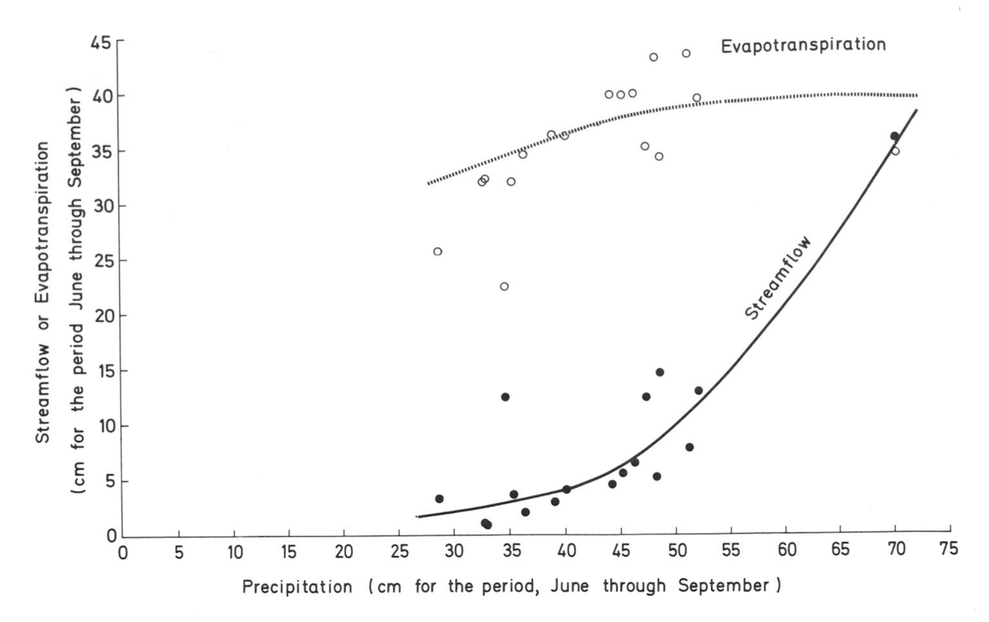

FIGURE 8. Relationship among precipitation, streamflow, and evapotranspiration during the months of June through September for Watershed 3 of the Hubbard Brook Experimental Forest during 1958–1974. Evapotranspiration (open circles) is determined as the difference between the amount of precipitation and streamflow for 1 June to 30 September. The solid line is fitted by eye. The dashed line represents evapotranspiration as computed from data on precipitation and air temperature using a U.S.D.A. Forest Service hydrologic model, which includes an estimate of the change in total storage of soil water. C.A. Federer, personal communication (1976).

downhill toward the stream channel. Frozen conditions in the stream beds add still further to the control of particulate matter losses in stream water as erodability in winter months normally is less than that during the growing season (Bormann et al., 1974).

Deep Seepage

One of the key features of the HBEF that enables a close accounting of water budgets is the watertight bedrock. Precipitation entering a watershed must leave by three avenues: (1) evaporation from leaf and soil surfaces, or both, to the atmosphere; (2) deep seepage, including lateral transfer through soil and bedrock across watershed boundaries and vertical movement through deep geologic strata; and (3) streamflow. Direct measurement of the first two is extremely difficult, if not impossible, with present technology and expertise. The latter, however, can be measured with fair degree of accuracy in many watersheds.

If there are strong indications that a given watershed has impermeable bedrock such that deep seepage is minimal and streamflow can be measured adequately, then the other major component of the water budget, evapotranspiration, can be derived quantitatively as the difference between streamflow and precipitation. Reasonable determinations of these components of the water budget for an entire water-year are feasible provided the water-year begins when storage of soil water is consistent each year.

There is substantial evidence that leads us to believe that the experimental watersheds at the HBEF have reasonable watertight bedrock and therefore have negligible water loss through deep seepage. The following is presented in support of this contention.

(I) From a report of a geological survey conducted at the HBEF, we highlight the findings of Bradley and Cushman (1956) regarding bedrock and watertightness:
   (a) Piezometric surface divides tend to coincide with the shape of the topography.
   (b) Unconsolidated deposits lie entirely within valleys which are rimmed by walls of bedrock.
   (c) The Littleton formation (an intensely metamorphosed rock, currently mapped as the Rangeley formation), is crossed by several sets of joints but openings are not large. These may extend one to a few meters below the surface but their size and number decrease with depth.
   (d) Only relatively small quantities of water occur in and move through these cracks.
   (e) They (Bradley and Cushman) concur with Fowler-Billings and Page (1942) and Moke (1946) that there is no evidence of significant faulting in the HBEF.
   (f) Solution cavities, such as may occur in limestone or marble, do not exist because the bedrock formations are not soluble.

Based on these findings, Bradley and Cushman conclude that it is improbable that measurable quantities of ground water move from one watershed to another or that measurable quantities of ground water escape from the watershed through major fault zones or solution openings.

(II) Assuming that two small adjacent Watersheds have reasonably similar precipitation, vegetative cover, soil, and evapotranspiration, then streamflow on a unit area basis should also be similar. If there is considerable deep seepage, however, then streamflow from the two watersheds is likely to be widely dissimilar. A comparison of streamflow among Watersheds 1, 3, 5, and 6 (11.8,

42.4, 21.9, and 13.2 ha, respectively) for 12 water-years was made using an analysis of variance. The forest on Watersheds 2 and 4 was cut in the late 1960s and therefore these two watersheds cannot be used in the long-term, simultaneous comparisons of streamflow with Watersheds 1, 3, 5, and 6.

In spite of wide variability in yearly streamflow during the period, e.g., 496 mm to 1,401 mm, there was no significant difference in streamflow among Watersheds 1, 3, 5, and 6. The calculated F value is 0.24, whereas the tabular F value at the 1% level is 4.26. Therefore, the probability that there are significant differences in streamflow between watersheds is extremely small. Because streamflow is virtually the same for each watershed, we conclude that the chances of deep seepage are negligible.

(III) As stated previously, if water loss through deep seepage were negligible, then evapotranspiration should approximate precipitation minus streamflow (runoff) for the common water balance equation: $ET = P - RO \pm \Delta S$, where P is precipitation, RO is streamflow, and ΔS is the change in storage of soil water. Over the course of an entire year we assume that the change in storage of soil water is zero. Hart (1966) compared $P - RO$ values for Watersheds 1 and 3 (HBEF) for 5 yr with calculated values of potential evapotranspiration based on the method by Thornthwaite (1948). Potential evapotranspiration, PE, is defined as the evaporation that would occur from a large fully wetted surface; that is, one with an unlimited water supply (Sellers, 1965). The average water-year values for Watersheds 1 and 3 for a 5-yr period (1958–1963) were 56.2 cm for PE and 53.2 cm for $P - RO$. There are several empirical methods for estimating PE (Hamon, 1961; Penman, 1956; Thornthwaite, 1948) and PE calculations by each of these vary slightly; in general however, such values of PE for HBEF are reasonably close to 56 cm. Actual evaportanspiration should be slightly less than PE except in rare cases. For practical purposes the values of $P - RO$ at the HBEF can be considered as close approximations of actual evapotranspiration. The remarkable closeness of the values of $P - RO$ and PE indicate that water loss from deep seepage cannot be a significant factor in the water balance at the HBEF.

(IV) A final bit of evidence that argues for the assumption of watertightness is the chloride budget. As far as we know, chloride is not accumulated in sizeable quantities within these watershed ecosystems in living or dead biomass or in the mineral soil. Moreover, chloride-bearing rocks in the HBEF are virtually nonexistent. Therefore, if losses by deep seepage are negligible, then inputs in precipitation should be observed quantitatively in the stream-water outputs from the ecosystem. As is shown in latter sections, on a long-term basis (9 yr) outputs of chloride have not been significantly different from inputs at the 0.05 probability level.

Representativeness of the Hydrology of Hubbard Brook for Northern New England

One of the main concerns about small experimental watersheds is whether the research results obtained can be interpreted and extended to other, larger watersheds. Climate, soil, geologic formation, topography, vegetative cover, and past land use all vary from place to place, yet they play important roles in the behavior of water fluxes through a watershed. In light of the influences of these watershed features on waterflow, the question arises: How representative is the hydrology of the HBEF to other areas of northern New England? Fortunately, there is one watershed characteristic that integrates the influences of most landscape features into a single measurable characteristic—streamflow. Considerable insight into the inner workings of a watershed ecosystem can be obtained by continuously monitoring the flow from that watershed.

In a comprehensive study of streamflow characteristics of watersheds in northern New England, Sopper and Lull (1965, 1970) addressed the question of representativeness of small experimental watersheds. They compared streamflow from eight watersheds in Maine, New Hampshire, Vermont, and Massachusetts with Watershed 1 at the HBEF.

Data adapted from Sopper and Lull are presented in Table 3. The eight watersheds were chosen on the basis of being less than 259 km^2 (100 mi^2) in size, being 90% or greater forest covered, having streams that were undisturbed by diversion or regulation, and having good to excellent records of precipitation and streamflow—the latter having simultaneous periods of records from 1 December 1956 to 30 November 1962. Annual and seasonal streamflow at the HBEF was very close to the average of the eight regional watersheds and fell within the maximum and minimum values for all periods.

Direct streamflow (that portion of the annual streamflow that rapidly leaves the watershed during and after storms, as opposed to prolonged base flow) for Hubbard Brook was high, as expected, but was less than direct streamflow from two other watersheds, one of which was much larger in size, 249 km^2.

Additional comparisons made by Sopper and Lull (Table 3) indicated close similarities in streamflow characteristics between experimental and regional watersheds, such as (1) number and amount of mean daily discharge above a given volume of flow and (2) flow duration (percent of time flow was equalled or exceeded).

TABLE 3. Hydrologic Characteristics for Selected Watersheds in Northern New England.[a]

Stream and location	Area km²	Forest cover %	Precipitation cm	Average annual and seasonal water yield, cm							Direct runoff, % of annual
				Annual	Spring	Summer	Fall	Winter	Growing season	Dormant season	
Austin Stream, ME	236	95	120.7	72.3	36.0	9.4	14.8	12.1	28.5	43.8	16.4
Swift River, ME	249	97	126.0	72.0	37.3	9.2	15.9	9.6	33.1	38.9	22.5
Ashuelot River, NH	184	94	113.2	58.4	32.7	3.6	10.5	11.6	15.7	42.7	11.0
N. Br. Contoocook R., NH	142	95	129.8	66.8	34.0	3.9	13.8	15.1	18.7	48.1	8.5
Otter Brook, NH	109	93	114.2	57.7	31.5	3.9	10.5	11.8	15.6	42.1	8.4
Smith River, NH	223	95	119.0	56.3	32.1	4.6	9.1	10.5	16.6	39.7	11.8
Saxtons River, VT	186	92	118.1	52.2	30.5	3.4	8.2	10.1	13.2	39.0	14.4
N. Br. Hoosic River, MA	104	90	155.5	74.8	38.9	6.3	13.2	16.4	20.6	54.2	23.8
Regional average	179	94	124.5	63.8	34.1	5.5	12.0	12.2	20.2	43.6	14.6
Hubbard Brook #1, NH	0.12	100	124.6	70.4	36.9	3.5	15.3	14.7	16.7	53.7	21.3

[a]After Sopper and Lull (1965, 1970).

In general, streamflow in small watersheds tends to have sharper peaks, higher storm levels, and shorter periods of sustained flow than it does in large watersheds. Such streamflow characteristics in small watersheds may be accentuated in steep, mountainous terrain with shallow to bedrock soils. Watersheds at the HBEF have these characteristics, but it is remarkable and reassuring that streamflow at HBEF, e.g., in a watershed of 0.12 km^2, closely approximates flow from watersheds that are more than three orders of magnitude greater in size!

The close relationships between stream-water quantity for streams in the HBEF and larger streams throughout northern New England add considerable support to the contention that HBEF streams are representative in many ways and that research results from the HBEF can be useful in extension and application to other areas for management purposes.

Addendum

The very dry year of 1964–1965 and wet year of 1973–1974 still remain the extremes in our long-term record (see Federer et al., 1990). The mean annual precipitation for W6 during 1963–1993 was 140.3 cm ($s_{\bar{x}}$ = 34.4) and annual streamflow was 87.9 cm, ($s_{\bar{x}}$ = 32.0). By difference the long-term mean evapotranspiration was 52.4 cm/yr, or 37% of the annual precipitation input. Transpiration accounts for about 60–70% of the evapotranspirational water loss on an annual basis at Hubbard Brook (C.A. Federer, personal communication). Obviously, the vast majority of the transpiration for this northern hardwood forest ecosystem occurs during the growing season from about 15 May to 15 September. This relation was demonstrated experimentally by a clearcutting experiment (see Bormann and Likens, 1979). The relationship between annual precipitation input, streamflow, and evapotranspiration (Figure 7) continues to be quite robust in the long-term record.

Some differences in hydrology exist between the experimental watersheds (W1–W6) largely related to east–west orientation. On average about 85 mm/yr or 6.4% more precipitation falls on the westernmost experimental watershed, W6, in comparison with the most easterly watershed, W3 (see Figure 2; Federer et al., 1990). Overall, however, differences between these south-facing watersheds of the HBEF are small (Table 4).

Based on the long-term record, we have no basis to question our assumption about the relatively water-tight nature of the experimental watershed-ecosystems of the HBEF. This feature of these watershed-ecosystems is of great importance for calculating quantitative, mass balances of water and chemicals. Certainly some small amount of deep seepage must occur, but is negligible relative to these balances. Currently,

TABLE 4. Annual Hydrologic Budget for Four Watershed-Ecosystems of the Hubbard Brook Experimental Forest During 1979–1980.[a]

Watershed-ecosystem	Area (ha)	Precipitation (cm)	Streamflow (cm)	Evapotranspiration[b] (cm)
W1	11.8	113.7	71.4	42.4
W3	42.4	113.9	69.3	44.6
W5	21.9	115.5	70.2	45.3
W6	13.2	119.3	70.8	48.5
Mean		115.63	70.41	45.21
Standard error of the mean		1.30	0.45	1.27
% of the total		100	61	39

[a]Modified from Likens et al. (1985).

[b]Precipitation minus streamflow; such estimates of annual evapotranspiration do not consider differences in soil water storage from year to year. Estimates of such storage differences between years, based on the BROOK model (Federer and Lash 1978), rarely exceed 4 cm and commonly are less than 1 cm. Neglecting changes in soil–water storage could lead to errors of up to 7% in estimating annual evapotranspiration.

detailed and extensive studies of groundwater flow are being done in the watershed for Mirror Lake (e.g., Paillet et al., 1987; Winter, 1985; Winter et al., 1989). This watershed, at lower elevation in the Hubbard Brook Valley and with deeper soils, has somewhat different hydrologic characteristics.

The actual path that water follows as it flows through the soil and/or downslope is of great importance biogeochemically. Much effort during the past decade or so has been expended at Hubbard Brook to quantify these relationships. For example, Stresky (1991) found that macropores, or "pipes," were common in the soils of the experimental watershed-ecosystems at Hubbard Brook. He found that the average macropore was 2.5 cm in diameter (0.5 to 7.5 cm), was 16 cm deep in the soil (2 to 130 cm), but comprised only about 0.2% of the area of soil profiles examined. Patterns of hillslope hydrology have been studied recently at Hubbard Brook by Shattuck (1991) and Mau (1993). Lawrence and Driscoll (1990) suggested that a combination of factors, including spatial variability in soil type and depth, modification of soil profiles adjacent to stream channels because of lateral flow patterns, macropore flow and vegetation type, determine the spatial and temporal patterns of stream-water chemistry observed at Hubbard Brook.

3 Chemistry

Weekly samples of precipitation and stream water are obtained from the experimental watersheds for chemical analysis. Rain and snow are sampled with continuously open plastic collectors (Likens et al., 1967). Samples of stream water are collected approximately 10 m upstream from each weir in clean polyethylene bottles. This is necessary because water collected from the ponding basin above the weir would be contaminated by calcium and bicarbonate from the cement in the weir itself. The concentrations of dissolved chemicals characterizing a period of time are reported as weighted averages. These averages are computed by summing the amount of chemical from individual samples of precipitation or stream water during the period and then dividing this value by the total amount of water during the period. Nutrient flux across ecosystem boundaries is determined (1) by multiplying the measured concentration of dissolved chemicals in the accumulated composite sample of precipitation by the amounts of precipitation during the interval and (2) by multiplying the average of measured concentrations of dissolved chemicals from stream water samples taken at the beginning and at the end of the interval by the amount of stream water during the period.

It has been shown that sampling according to a standard time series may seriously underestimate or overestimate a highly variable parameter (e.g., Claridge, 1970); however, most of the dissolved chemicals in stream water at Hubbard Brook fluctuate relatively little in concentration so that serious errors are not produced by weekly or even monthly sampling in some cases (Johnson et al., 1969; Likens et al., 1967).

As standard practice, we do not include data from precipitation samples that contain leaves, bud scales, insects, bird feces, or any other obvious foreign particles. Such contamination represents a serious problem, for a variety of substances may leach from these materials as the water comes in contact with them in the collector. Our intent is to have enough collectors in position to provide at least one "clean" sample per interval. This procedure normally works very well.

Details concerning the routine methods used in collecting samples of precipitation and stream water and analytical procedures have been given by Bormann and Likens (1967), Likens et al. (1967), Fisher et al. (1968), Bormann et al. (1969), and Hobbie and Likens (1973).

Precipitation Chemistry

Precipitation chemistry has been monitored at the HBEF since 1963. To our knowledge this unbroken series is the longest comprehensive record of precipitation chemistry in the United States. Moreover, samples are collected on a weekly or storm basis instead of on the more commonly used monthly interval. Considering the problems of contamination and biogeochemical transformations that can occur in the reservoir of a bulk precipitation collector, experience has shown that sampling intervals of not longer than a week are highly desirable, if not necessary, to obtain accurate data on precipitation chemistry.

Annual Weighted Concentrations

The chemical concentration of precipitation is dominated by H^+ ions which constitute 70% of the total cationic strength. The remaining 30% is shared among NH_4^+, Ca^{2+}, Na^+, Mg^{2+}, and K^+ in that order (Table 5).

Sulfate and hydrogen ion are the most prevalent inorganic ions in bulk precipitation at Hubbard Brook (Table 5). On an equivalent basis sulfate is 2.5 times more important than the next most abundant anion, NO_3^-, hydrogen ion is 5.9 times more abundant than the next most abundant cation, NH_4^+; in both cases these ions far exceed the total equivalency for all other anions and cations. Essentially, then, the incident precipitation at Hubbard Brook may be characterized as a solution of sulfuric and nitric acids at a pH of about 4.1. On a long-term basis the total negative equivalent value is not statistically different ($p < 0.05$) from the total positive value (Tables 5 and 8).

Surprising concentrations of dissolved organic carbon are found in rain and snow at Hubbard Brook (Table 5). Precipitation falling on the Hubbard Brook watersheds contains an average total organic carbon concentration of 1.28 mg/l of which 84% was dissolved (Likens et al., 1983). Only a small fraction of this organic matter is dissociated organic acids (Likens et al., 1976a, 1983; Galloway et al., 1976). Particulate plus dissolved macromolecular (>1000 MW) organics accounted for 51%, and carboxylic acids represented 14%, aldehydes 11%, carbohydrates 8%, and tannin/lignin 8% of the total organic carbon in precipitation during 1976–1977 (Likens et al., 1983). The actual composition of the dissolved organic carbon in precipitation is probably a complex mixture of many organic species, most of which are present in trace amounts. Dissolved

TABLE 5. Weighted Annual Mean Concentration of Dissolved Substances in Bulk Precipitation and Stream Water for Undisturbed Watersheds 1–6 of the Hubbard Brook Experimental Forest.

	Precipitation 1963–1974		Stream water 1963–1974	
Substance	mg/l	μEq/l	mg/l	μEq/l
H^+	0.073[a,b]	72.4	0.012[a,c]	11.9
NH_4^+	0.22[b]	12.2	0.04[b]	2.2
Ca^{2+}	0.16	7.98	1.65	82.3
Na^+	0.12	5.22	0.87	37.8
Mg^{2+}	0.04	3.29	0.38	31.3
K^+	0.07	1.79	0.23	5.9
Al^{3+}	—[d]	—	0.24[e]	26.6
SO_4^{2-}	2.9[b]	60.3	6.3[b]	131
NO_3^-	1.47[b]	23.7	2.01[b]	32.4
Cl^-	0.47[c]	13.3	0.55[c]	15.5
PO_4^{3-}	0.008[f]	0.253	0.0023[g]	0.1
HCO_3^-	~0.006[h]	0.098	0.92[i]	15.1
Dissolved silica	—[d]	—	4.5[b]	—
Dissolved organic carbon	1.1[j]	—	1.0[k]	—
pH	4.14		4.92	
Total	8.64	(+) 102.9 (−) 97.7	18.70	(+) 198.0 (−) 194.1

[a]Calculated from weekly measurements of pH.
[b]1964–1974.
[c]1965–1974.
[d]Not determined, trace quantities.
[e]1964–1970.
[f]1972–1974.
[g]1967–1968, 1972–1974.
[h]Calculated from H^+–HCO_3^- equilibrium.
[i]Watershed 4 only, 1965–1970.
[j]1976–1977 (Likens et al., 1983)
[k]Watershed 6 only, 1967–1969.

carbon in precipitation represents an additional energy source for the ecosystem and may be of some ecological importance. Studies of these relationships are currently underway at Hubbard Brook.

Origin of Ions in Precipitation

Cations and anions in precipitation originate from a variety of sources, including oceanic spray, terrestrial dust, gaseous pollutants, and volcanic emissions. How ionic materials find their way into a precipitation collector is an important enigma. The solubility of atmospheric gases, from either natural and anthropogenic origin, in atmospheric water is obvious but the origin of such nonvolatile constituents as Na, Mg, Ca, and K in precipitation is not so obvious. The dispersion of sea salts, such as NaCl and $MgSO_4$, in atmospheric aerosols is one fairly well-known process (Junge,

1963; MacIntyre, 1974). Nevertheless, only a part of the Ca and K in precipitation can be reasonably attributed to marine sources and the "excess" is customarily assigned to unknown "continental sources" (Cogbill and Likens, 1974; Junge, 1963; Granat, 1972). At Hubbard Brook the long-term ratios for ionic Na:Cl are roughly double those anticipated from transported sea salt (cf. Table 5; Juang and Johnson, 1967; Junge, 1963). In contrast, the ionic Ca:Mg ratio is representative of continental precipitation (Eaton et al., 1973; Ericksson, 1952).

One explanation for the excess cations in precipitation is that soil particles and aerosols in the atmosphere may be chemically altered to produce cations. The ubiquitous hydrogen ions in precipitation may react with the entrained soil particles and release cations into solution (see subsequent sections). Therefore, some of the cations found in solution in a precipitation collector may have been derived from an "artificial weathering" reaction within the continuously open (bulk) collector such as we use at Hubbard Brook. Such "weathering reactions" that occur within the precipitation collection reservoirs would not change the total amount of cations input to the ecosystem but might alter their chemical form or underestimate the input of hydrogen ion in precipitation.

Cations and anions in precipitation are approximately balanced. Small errors (~0.05 pH unit) in the determination of pH, which has been used to estimate the concentration of hydrogen ion, are sufficient to explain the discrepancy in the cation–anion balance (Table 5). Considering that these long-term averages include the various sampling and analytical errors over an 8- to 11-yr span, the agreement obtained in equivalents is quite good.

It is important relative to quantitative meteorologic inputs to determine whether the dissolved and particulate materials in a bulk precipitation sample have originated outside the ecosystem's boundaries or whether they have merely been circulating within the boundaries. Because the Hubbard Brook watersheds and the surrounding region (1) are almost entirely forested or have other vegetational ground cover, (2) have a long period of snow cover, and (3) have generally humid conditions during other seasons, we believe that relatively little dust arises from within the ecosystem and that our bulk precipitation collectors measure dissolved and particulate matter that has largely originated outside the ecosystem's boundaries.

Bulk precipitation collectors may be very inefficient in collecting dry deposition, particularly aerosols smaller than 1 μm. Preliminary estimates, based on some measurements of aerosol concentrations and a reasonable deposition velocity, suggest that the dry deposition of Ca, Mg, Na, K, and Cl is probably a small fraction of bulk precipitation inputs at Hubbard Brook. This is not necessarily the case for nitrogen and sulfur. Small aerosol particles and gases of nitrogen and sulfur may be generated from biogenic activity and from the combustion of fossil fuels. Both gases and small aerosols may be transported long distances in the atmosphere. These chemicals therefore exist in the atmosphere at Hubbard Brook in much higher concentrations than that expected from the local environment, and their dry deposition is underestimated by bulk precipitation collectors. This matter is considered more fully in later sections.

Elevational Effects

There is no significant difference in the content of Ca^{2+}, Mg^{2+}, Na^+, or K^+ in precipitation at different elevations [from 545 to 758 m MSL (mean sea level)] within the experimental watersheds (Likens et al., 1967). Furthermore, a study done in 1971–1972 indicated no significant difference between the content of Ca^{2+}, Mg^{2+}, Na^+, K^+, NH_4^+, or Cl^- in precipitation samples from the experimental watersheds at 610 m MSL and those from the Forest Service Headquarters Station at 252 m MSL. However, based upon an analysis of paired samples from these two locations, the concentrations of SO_4^{2-} and NO_3^- were significantly higher ($p < 1\%$ and $p < 5\%$, respectively) at the lower station. The source of this variation may be related to human activity at the lower elevations but this has not been confirmed.

Acid Precipitation

One of the more interesting and ecologically significant findings from these studies was the high acidity of rain and snow at Hubbard Brook (Likens et al., 1972). The average annual weighted pH from 1964–1965 through 1973–1974 ranged between 4.03 and 4.21 (Figure 9); the lowest value recorded for a storm at Hubbard Brook was pH 2.85 and the highest value was 5.95. During the years 1969–1974 of the study only, no value for weekly precipitation exceeded a pH value of 5.0. Such precipitation is decidedly abnormal chemically because pure water in equilibrium with atmospheric concentrations of CO_2 should have a pH of not less than about 5.6 (Barrett and Brodin, 1955). In other words, the precipitation at Hubbard Brook has a hydrogen ion concentration 50–500 times greater than expected. The increased acidity of precipitation is the result of anthropogenic emissions of SO_2 and NO_x, which are hydrolyzed and oxidized to strong acids (H_2SO_4, HNO_3) in the atmosphere (Bolin, 1971; Likens and Bormann, 1974b; Likens et al., 1972). The occurrence of acid precipitation at Hubbard Brook, which is >100 km distant from any large urban-industrial area, emphasizes that one of the major ways humans can influence natural ecosystems is through pollution of the atmosphere and that these effects are not limited to areas adjacent to sources of pollution.

Acid precipitation apparently has been falling on much of the eastern United States since at least the early 1950s (Cogbill and Likens, 1974;

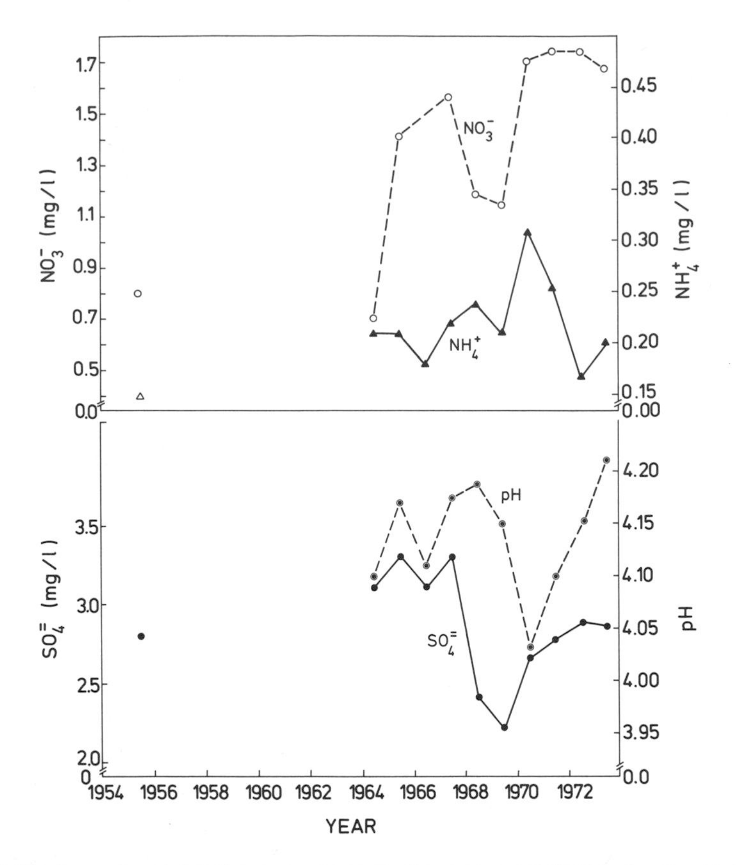

FIGURE 9. Annual weighted mean concentrations in precipitation for the Hubbard Brook Experimental Forest during 1955–1974. The values for 1955–1956 were extrapolated from isopleth maps given by Junge (1958) and Junge and Werby (1958). Note that the ordinate has been compressed. From Likens and Bormann (1974b).

Likens and Bormann, 1974b). The record for pH of precipitation at Hubbard Brook, although short, is the longest in the United States and provides some information on trends. There has been a slight upward trend in annual weighted average concentrations of hydrogen ion between 1964–1965 and 1970–1971, followed by a downward trend until 1973–1974; overall (1964–1974), however, no trend in concentration is statistically significant (Figure 9). Likewise, concentrations for SO_4^{2-} and NH_4^+ vary from year to year but there are no statistically significant trends for the period. In contrast, annual weighted NO_3^- concentrations currently are about 2.3-fold greater than they were in 1955–1956 or in 1964–1965 (Figure 9). An even larger relative increase in nitrate concentration was observed in the central Finger Lakes Region since 1945 (Likens, 1972).

These increases in nitrate concentrations are roughly correlated with the increased use of fossil fuels in combustion engines (Likens, 1972) and with increased use of nitrogenous fertilizers, but there may be other explanations. Nevertheless, the increase in nitrate concentration in precipitation in this remote, forested area at Hubbard Brook is indicative of an increasing and widespread distribution of atmospheric pollutants. The ecological implications are difficult to judge. On the one hand, increased nitrate input into the northern hardwood forest ecosystem and associated aquatic ecosystems may act as nitrogen fertilizer; on the other hand, nitrate contributes to the formation of a potentially harmful increase in acidity of precipitation.

Even though the hydrogen ion concentrations are variable from year to year, the annual input (concentration times volume) in precipitation increased by 1.4-fold during the period 1964–1965 to 1973–1974 (Figure 10). This increased input of hydrogen ion was in sharp contrast to the annual input of all other ions, which remained constant or decreased, except nitrate (Table 6). Based upon a regression analysis, annual nitrate input increased by 2.3-fold during the decade; there was no significant increase in annual sulfate input during the period (Likens et al., 1976b).

The increased annual input of hydrogen ion is partially explained by the "increase" in the amount of annual precipitation during the 10-yr period (Table 6); that is, the early years of the decade (1963–1965) were drought years and the last year (1973–1974) was extremely wet (Figure 4). However, close examination of data for individual years shows that factors other than increased amount of precipitation are important

TABLE 6. Dependence of Annual Precipitation Input on Sequential Years for the Hubbard Brook Experimental Forest.[a]

Substance	Slope	Correlation coefficient	Time period
Ca^{2+}	−0.174[b]	−0.83	1963–1974
Mg^{2+}	−0.046[c]	−0.66	1963–1974
K^{+}	−0.126[c]	−0.66	1963–1974
Na^{+}	−0.002	−0.02	1963–1974
NH^{+}	0.109	0.57	1964–1974
H^{+}	0.033[c]	0.73	1964–1974
SO_4^{2-}	0.838	0.32	1964–1974
NO_3^{-}	1.73[b]	0.79	1964–1974
Cl^{-}	0.300	0.22	1967–1974
Water	4.95[c]	0.72	1963–1974

[a]As determined by regression analysis when $Y = ax + b$, where Y = input in kg of dissolved substances/ha or cm of water per unit area, a = slope, x = water-year (x_1 = 1963–1964, x_2 = 1964–1965, . . ., x_{11} = 1973–1974) and b = Y intercept.

[b]Probability of a larger F value < 0.01.

[c]Probability of a larger F value < 0.05.

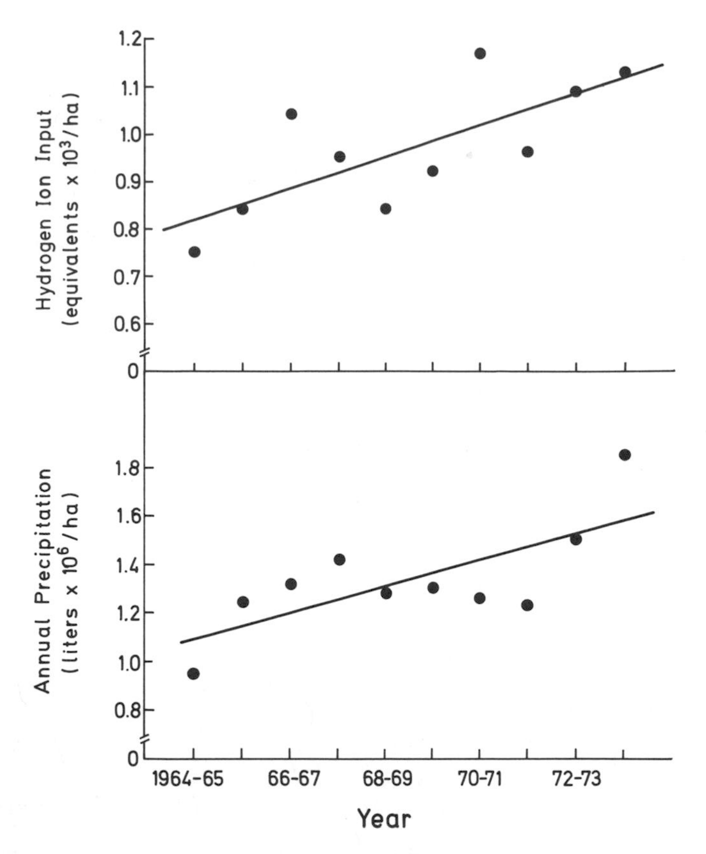

FIGURE 10. Annual hydrogen ion and precipitation input for the Hubbard Brook Experimental Forest. The regression line for hydrogen ion is $Y = 0.033x + 0.819$ where Y is the H^+ input in 10^3 Eq/ha, and x is the year; the significant correlation coefficient is 0.73. Note that the ordinate has been compressed. From Likens et al. (1976b).

(Figure 10). Consequently, annual concentrations alone, even though they are weighted for volume of precipitation, do not accurately reflect trends in the annual input of hydrogen ion.

Sulfuric acid is the dominant acid in precipitation at Hubbard Brook. Based on a stoichiometric formation process in which a sea salt anionic component is subtracted from the total anions (Cogbill and Likens, 1974), sulfate contributed more than 65% of the acidity in precipitation throughout the decade. However neither the concentration nor the input of sulfate changed significantly during the period of our study (Figure 9 and Table 6). Therefore it is not surprising that the increase in input of hydrogen ion from 1964 to 1974 is not significantly related to the change in input of sulfate. In contrast, the increase in input of hydrogen ion was highly correlated with the increased input of nitrate (Figure 11). The relationship between annual inputs of hydrogen ion and nitrate is a powerful argument that nitric acid is the crucial variable for explaining the increased input of hydrogen ion during the past 10 yr (Likens et al., 1976b). Although sulfuric acid dominates the precipitation at Hubbard Brook and accounts for most of the hydrogen ion in precipitation, therefore, the increase in input of hydrogen ion during the past decade apparently has been caused by an increase in the annual nitric acid content of precipitation falling on this forested ecosystem.

Precipitation chemistry has changed both qualitatively and quantitatively at Hubbard Brook during the past decade. Absolute concentrations have varied (Figure 9) and relative proportions of the component chemicals have changed (Figure 12). Based upon stoichiometry, the sulfate contribution to acidity dropped from 83 to 66% and nitrate increased from 15 to 30% from 1964–1965 to 1973–1974. Annual inputs reflect these changes in complex ways. In a further attempt to resolve the relative importance of the various factors that controlled the annual hydrogen ion inputs, a stepwise multiple regression analysis was done to relate the

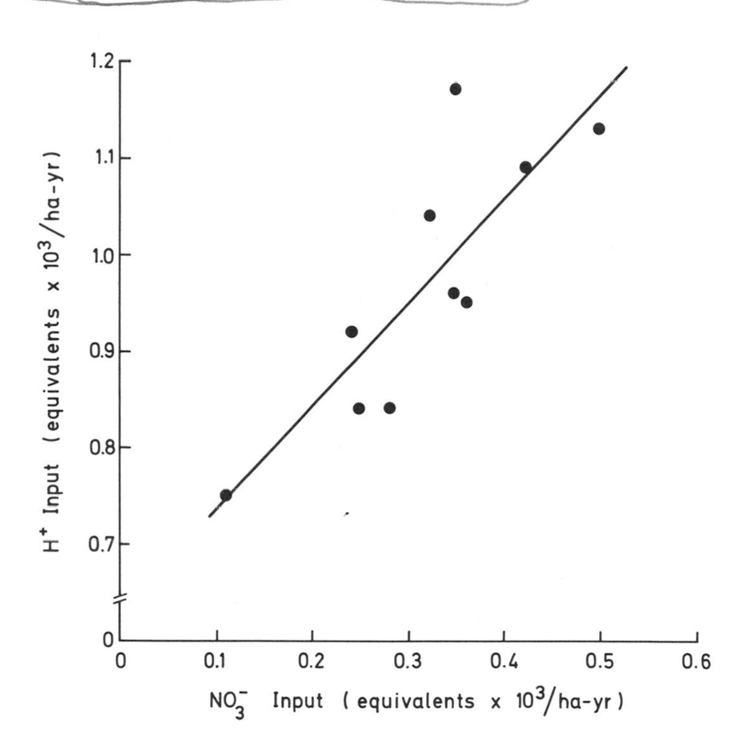

FIGURE 11. Relationship between the annual hydrogen ion input and the annual nitrate input during the period 1964–1965 to 1973–1974. The regression line, $Y = 1.07x + 0.631$, where Y is annual input of hydrogen ion in 10^3 Eq/ha · yr and x is annual nitrate input in 10^3 Eq/ha · yr, is highly significant (correlation coefficient of 0.84, and probability of a larger F value is <0.01). From Likens et al. (1976b).

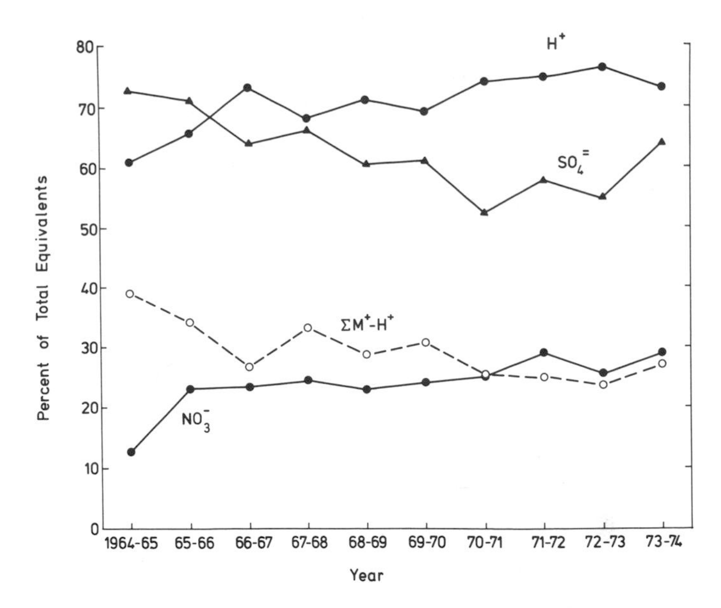

FIGURE 12. Variation in ionic composition of precipitation for the Hubbard Brook Experimental Forest during 1964–1965 to 1973–1974. From Likens et al. (1976b).

annual hydrogen ion input to a variety of independent variables. An analysis of five independent variables indicated that 86% of the variability in annual hydrogen ion input during the decade was related to annual nitrate input. Six percent of the variability was caused by annual sulfate input, 5% by the input of the sum of all cations minus hydrogen ion, 2% by year, and $<0.01\%$ by the annual amount of precipitation.

These data illustrate how economic decisions and activities far removed from this region may influence the dynamics of the northern hardwood ecosystem. Apparently several decades ago, changes in atmospheric chemistry resulting from changes in fossil fuel consumption and disposal of gaseous effluents resulted in decreased pH of precipitation over the northeastern United Sates. Initially, changes in H^+ concentration were related to changes in sulfur chemistry (Likens et al., 1972) but during the last decade, although sulfuric acid dominated the precipitation at Hubbard Brook, increased input of H^+ was caused by an increase in atmospheric NO_3^-.

The biogeochemical effects of excess acidity in precipitation and its increased input during the past decade at Hubbard Brook are currently under study. Because precipitation is weakly buffered, the presence of small amounts of strong mineral acids, such as H_2SO_4, can greatly depress the pH value. Even though only relatively small amounts of acid are involved the potential effect of increased amounts of hydrogen ions on biologic and chemical systems can be large.

Fish kills of major proportions and diminished reproduction of salmonoid fish have been observed and attributed to acid precipitation in areas of similar geology in the Adirondack Mountains of New York (Schofield, 1976), in Canada (Beamish, 1976), and in Scandinavia (Wright et al., 1976). No fish are found in the small headwater streams at Hubbard Brook, but the effect of the acidic precipitation on this distribution is not known. Eastern brook trout occur in the streams at lower elevations in the HBEF.

The first portions of meltwater (~20%) from the snowpack may be significantly more acid than subsequent portions (Hornbeck et al., 1976; Johannessen et al., 1976). This phenomenon may cause a sharp drop in the pH of drainage streams and lakes during the spring and has been blamed for massive fish kills in Scandinavia (Wright et al., 1976). This seasonal change in pH also has been observed in streams and in Mirror Lake within the Hubbard Brook Valley, but it is a very temporal condition of variable intensity. During the winter and with intermittent melting, the snowpack becomes less acid than the incident precipitation (Hornbeck et al., 1976). If the snow melts rapidly there may be a flush of highly acidic water into the stream channels; however, if the snowpack melts more slowly the reduction of pH in stream water is much less pronounced.

Despite seasonal differences in acidity of precipitation, effects of snow accumulation and melt, and variations in residence time of soil water, the acidity of stream water at Hubbard Brook at the level of the gaging weirs is remarkably constant throughout the year (Figure 13). Even though the average monthly concentration for hydrogen ion in precipitation for Watershed 6 ranged from 46 to 102 μEq/l, average monthly concentrations in stream water from Watershed 6 only varied from 10 to 17 μEq/l during 1964–1974. The uniformity of stream-water acidity indicates a strong buffering action of the terrestrial ecosystem (Figure 14). The terrestrial ecosystem is therefore very effective in ameliorating the short-term impact of the acid precipitation on the associated aquatic ecosystems.

Our model of the forest ecosystem (Figure 1) suggests a number of functions internal to the forested ecosystem that may be affected by acid precipitation. For example, leaching of substances from the canopy can be speeded up and, in fact, that seems to be the case. Eaton et al. (1973) found that 90% of the H^+ ions striking the summer canopy of the forest were consumed within the canopy, presumably releasing an equivalent amount of basic cations. Laboratory studies of seedlings of major tree species at Hubbard Brook (Wood and Bormann, 1974, 1975) indicated increased leaching of cations as the H^+ content of artificially applied mist increased.

The podzol soils at Hubbard Brook are already quite acid and the geochemical effects of acid precipitation appear to be minimal (Johnson et al., 1972). However, over the long term, acid precipitation could have an important effect on a variety of soil processes, e.g., nitrogen fixation

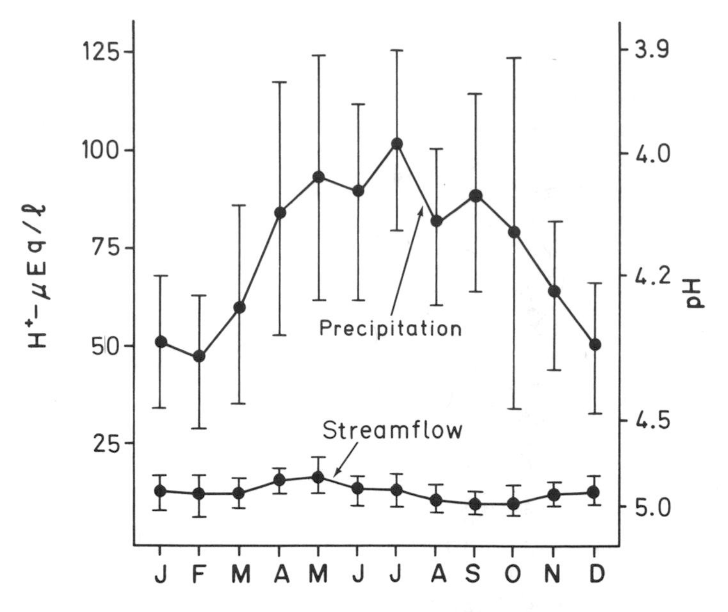

FIGURE 13. Weighted monthly hydrogen ion concentration and pH for precipitation and streamflow for Watershed 6 of the Hubbard Brook Experimental Forest during 1965–1974. The vertical lines for each month represent one standard deviation. From Hornbeck et al. (1976).

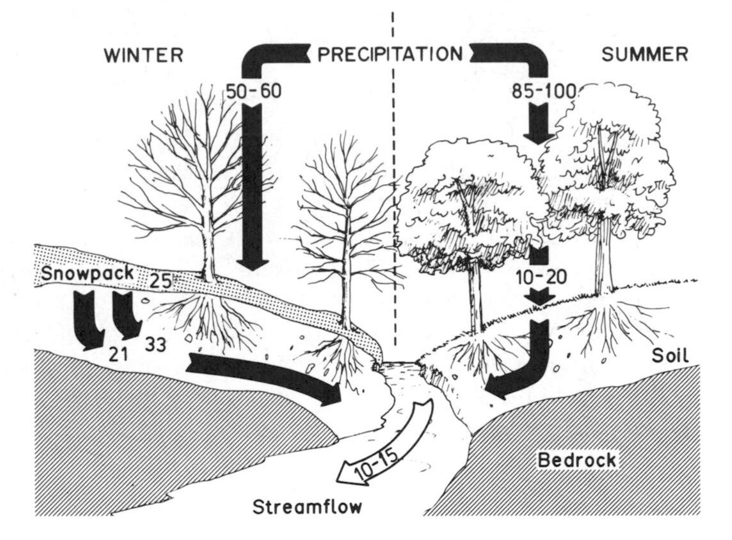

FIGURE 14. General relationships for two seasonal periods at Hubbard Brook. The values are hydrogen ion concentrations in μEq/liter. Modified from Hornbeck et al. (1976).

or nitrification. Theoretically 1 m of rain at pH 4 could leach 50 kg/ha of $CaCO_2$ from the system in drainage water.

Whittaker et al. (1974) and Cogbill (1976) reported a decline in forest growth at Hubbard Brook during the last two decades; in general, however, significant effects of acid rain on forest growth are extremely difficult to detect. The forest ecosystem is regulated by a large number of such environmental factors as the amount and distribution of rainfall, temperature, the length of growing season, unusual meteorologic events, the availability of nutrients, and insect and disease outbreaks. It is extremely difficult to separate the effects of these variables from those of acid precipitation. It is entirely possible, however, that acid rain may be reducing forest growth by small and persistent amounts over wide areas or upsetting the ecological balance so as to reduce the capacity of certain species to resist insect or disease attacks. Cumulative effects over a number of years may result in serious deterioration of the forest ecosystem, particularly as acid precipitation is but one of an array of pollutants affecting the landscape (Bormann, 1974).

Calculation of Mass Input of Chemicals

The meteorologic input of chemicals for the HBEF is calculated by multiplying the measured concentrations of each ion in the weekly precipitation sample (mg/l or mEq/l) by the volume of precipitation incident over the watersheds for that week (l/ha). The meteorologic input of ions is then expressed in units of mass (or chemical equivalency) per unit area per week. The weekly inputs then may be summed conveniently into monthly, seasonal, or yearly increments.

Meteorologic Input—Dissolved Substances

The average annual input of dissolved inorganic substances in rain and snow was calculated to be 72.5 kg/ha or 1,320 Eq/ha during 1963–1974 (see Table 11). The input of specific elements is dealt with in a later section when nutrient budgets are discussed. During 1964–1965, one of the drought years, the total input in equivalents was exceptionally low (23% less than the long-term average); during the wet year, 1973–1974, the ionic input was 24% greater than the long-term average (Figure 15). Overall the flux of chemicals in precipitation to these forested ecosystems is substantial, both as nutrients and as weathering agents. This is particularly true for H^+, NH_4^+, NO_3^-, Cl^-, PO_4^{3-}, and SO_4^{2-} and to a lesser extent for K^+, Ca^{2+}, Mg^{2+}, and Na^+.

Throughfall and Stemflow

The water that reaches the forest floor is of vastly different chemical content than the incident precipitation. Only about 87% of the incident

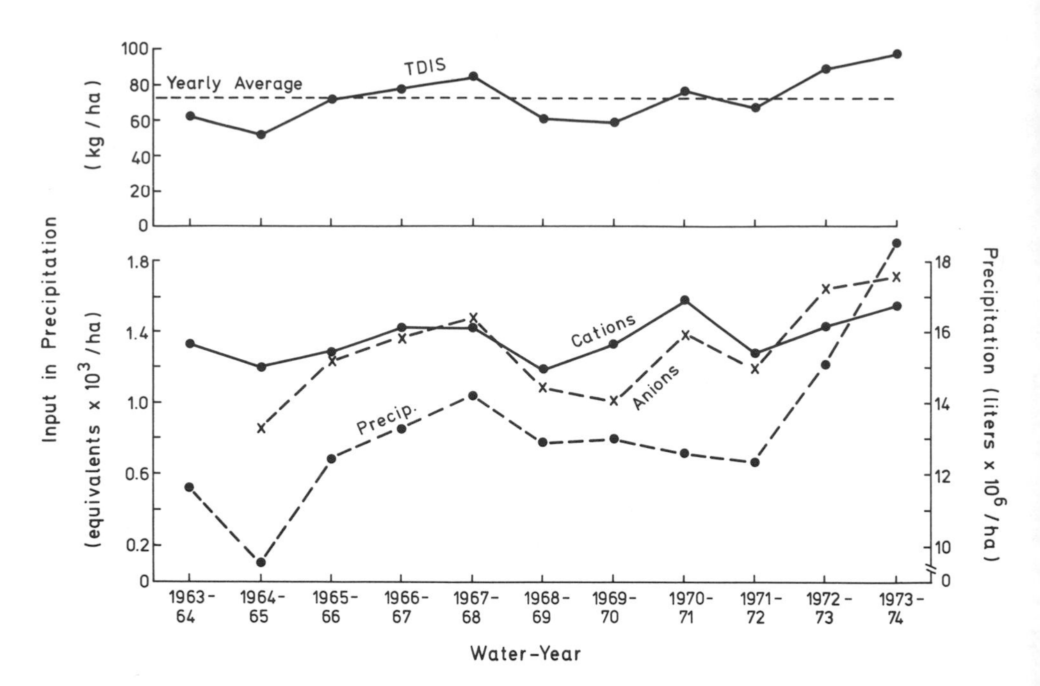

FIGURE 15. Average annual input of dissolved inorganic substances in precipitation to the experimental Watersheds 1–6 of Hubbard Brook Experimental Forest. TDIS = total dissolved inorganic substances.

precipitation actually reaches the ground of the forest ecosystem during the growing season. The remainder is evaporated (or absorbed) directly from the surface of leaves, branches, and stems (Leonard, 1961). As the precipitation passes through the forest canopy its chemistry is altered significantly (Table 7). Throughfall and stemflow (about 5% runs down the stems of trees) during the growing season are greatly enriched in potassium (about 91 times), phosphorus (18 times), magnesium (15 times), and calcium (10 times). All other dissolved substances also increase, with the exception of hydrogen ion, which is held within the canopy by cation exchange reactions.

Based on an average carbon concentration of 2.4 mg/l (1973–1974) and an average precipitation of 129.5×10^5 l/ha·yr, some 31 kg/ha·yr of dissolved organic carbon are added to the Hubbard Brook ecosystems each year in direct precipitation. As the precipitation passes through the forest canopy it is enriched in organic carbon (Table 7). Throughfall and stemflow during the growing season contain some 52 kg carbon per hectare (Eaton et al., 1973), so an additional 21 kg carbon per hectare must be leached from the forest canopy and added to that already present in the precipitation. These relationships probably are much more complicated than this, because the microflora of the leaf and bark surfaces of the

Table 7. Weighted Average Chemical Composition of Incident Precipitation and Throughfall Under Different Size Classes of Sugar Maple, Yellow Birch, and American Beech Trees During 1 June 1969 to 28 October 1969.[a]

Material	Precipitation above canopy	Precipitation under canopy (throughfall and stemflow)	Net change
Ca	0.16	1.59	+1.43
Mg	0.03	0.45	+0.42
K	0.07	6.37	+6.30
Na	0.06	0.14	+0.08
NO_3-N	0.22	0.67	+0.45
NH_4-N	0.21	1.21	+1.00
Total N	0.44	2.44	+2.00
PO_4-P	0.0026[b]	0.15	+0.147
SO_4-S	0.9	5.4	+4.5
Cl	0.45	1.46	+1.01
H	0.087	0.010	−0.077
Organic C	2.4[c]	12	+10

[a]Values are in mg/l (modified from Eaton et al., 1973).

[b]Weighted average for 1972–1974.

[c]Weighted average for 1973–1974.

vegetation may metabolize the more labile, dissolved organic substances in precipitation and in turn secrete other organic substances.

There are various sources for the dissolved substances in throughfall and stemflow, but their relative importance is not entirely clear (Eaton et al., 1973): (1) Some nutrients are contained in the incident precipitation; (2) some of the nutrients may have been impacted aerosols that are washed off by the incident precipitation. These two portions should be considered as a part of the chemical flux from outside the ecosystem. Similarly, (3) nutrients with a normal gaseous phase, which were incorporated directly on or into the plant (for example, fixation of C by photosynthesis, absorption, and reaction with SO_2 or NH_3) and then removed from tissues by incident precipitation should be considered as meteorologic inputs. The remaining nutrients in throughfall and stemflow, which have been leached from the vegetational tissues or associated microflora, are nutrients cycling primarily within the intrasystem cycle of the ecosystem (Figure 1) and should not be considered part of the meteorologic flux. We have not been able to quantify the amounts of chemicals transported along each of these pathways for the ecosystems at Hubbard Brook. Some specific data are presented later in the discussion of nutrient cycles at Hubbard Brook.

Stream-Water Chemistry

The chemistry of stream water has been monitored in Watersheds 1–6 of the HBEF since 1963. Experiments that involved cutting of the forest vegetation were undertaken in Watershed 2, starting in November of 1965, and in Watershed 4, starting in October of 1970. After these dates, therefore, data on stream-water chemistry from these two watersheds were not included in the results presented here.

Dissolved Substances

The Hubbard Brook forest ecosystem brings about both qualitative and quantitative chemical changes in rain and snowmelt water as it passes through the system (Table 4). Water enters the system as a dilute solution of sulfuric and nitric acid (pH ~ 4) but leaves the system containing primarily neutral sulfates and nitrates (pH ~ 5). Calcium and sulfate dominate the stream-water chemistry, but sodium, magnesium, and aluminum also are relatively important cations on an equivalent basis. Sulfate is four times more abundant than the next most abundant anion, NO_3^-, in stream water. The balance between cations and anions in stream water is very good over an 11-yr span (Tables 5 and 8). Annually the smaller value (on an equivalent basis) was always within 10% of the larger value.

The ionic strength of stream water (0.20 mEq/l) is twice that of the incoming water (0.10 mEq/l). A principal factor here is the concentration effect (distillation) of evapotranspiration. As water is lost from the system by evapotranspiration, chemicals in solution tend to be concentrated into a smaller volume of water. Based upon an average annual evapotranspiration loss of 38% (Table 1), the concentration factor would be 1.6, which would fall short of the observed concentration factor for chemicals (2.0). However, the concentration factor is not so straightforward as it appears, for not only do concentrations in stream water change but the proportions of ionic or dissolved species change as well, indicating important internal chemical and biologic reactions. In any case, an additional quantity of dissolved salts is acquired by the water from within the ecosystem's boundaries. This process is related to chemical weathering reactions (Johnson et al., 1968) and is discussed separately.

Appreciable amounts of dissolved organic matter are found in the undisturbed tributaries to Hubbard Brook. Fisher (1970) reported an average dissolved organic matter concentration of 2.34 mg/l during 1968–1969 in Bear Brook. This would be 1.05 mg of carbon per liter if the dissolved organic matter were 45% carbon. Hobbie and Likens (1973) found an average annual weighted concentration of 1.0 mg dissolved organic carbon per liter in stream water from Watershed 6. Most of the stream-water carbon values ranged between 0.3 and 2.0 mg/l, but a

TABLE 8. Arithmetic Mean of Annual Weighted Concentrations of Dissolved Substances in Bulk Precipitation and Stream Water for Undisturbed Watersheds 1–6 of the Hubbard Brook Experimental Forest During 1963–1974.

	Precipitation, mg/l				Stream water, mg/l			
Substance	$\bar{x}$	$s_{\bar{x}}$	n^a	$s_{\bar{x}}/\bar{x}$, %	$\bar{x}$	$s_{\bar{x}}$	n^a	$s_{\bar{x}}/\bar{x}$, %
Ca^{2+}	0.17	0.02	11	11.8	1.65	0.06	11	3.6
Mg^{2+}	0.05	0.01	11	20.0	0.38	0.01	11	2.6
K^+	0.07	0.02	11	28.6	0.23	0.01	11	4.3
Na^+	0.12	0.01	11	8.3	0.88	0.02	11	2.3
Al^{3+}	—	—	—	—	0.23	0.01	6	4.3
NH_4^+	0.22	0.01	10	4.5	0.04	0.01	10	25.0
H^+	0.0739	0.003	10	4.4	0.0126	0.0018	9	14.0
SO_4^{2-}	2.87	0.11	10	3.8	6.23	0.11	10	1.8
NO_3^-	1.43	0.11	10	7.7	1.93	0.31	10	16.1
Cl^-	0.51	0.09	7	17.6	0.54	0.02	7	3.7
PO_4^{3-}	0.008	—	2	—	0.002	—	2	—
HCO_3^-	—	—	—	—	1.62	0.12	5	13.5
Dissolved silica	—	—	—	—	4.43	0.11	10	2.5

[a]years of measurement.

maximum carbon value of 4.8 mg/l was observed during an extremely high streamflow period (29 July 1969, 34×10^4 l/ha·day).

In general the measured stream-water concentrations of most dissolved substances vary within a narrow range (less than a factor of two), even though discharge of water may fluctuate over four orders of magnitude during an annual cycle (Fisher and Likens, 1973; Johnson et al., 1969; Likens et al., 1967). This is particularly true for magnesium, sulfate, chloride, and calcium concentrations. Sodium and dissolved silica concentrations may be diluted up to threefold during periods of high streamflow, whereas aluminum, hydrogen ion, dissolved organic carbon, nitrate, and potassium concentrations are increased with increased discharge. Biotic activity within the ecosystem plays an important role in determining these relationships for potassium and nitrate in stream water. Nitrate and potassium are quite sensitive indicators of biologic activity; therefore, stream-water concentrations for these two nutrients are markedly reduced during periods of plant growth and increased during periods of vegetation dormancy.

To empirically describe these variations in stream-water chemistry we have developed a model that predicts stream-water concentration in relation to stream discharge (Johnson et al., 1969). The model is based on some rather simple and intelligible assumptions. First of all, we assume that two discrete water types exist in our watershed ecosystem; one is

PHOTOGRAPH 6. *Winter conditions at the HBEF.* Note that the snowpack has mostly buried the instrument shelter for the gaging weir on Watershed 6 (see page 21).

represented by relatively dilute water recently added to the system in rain or snowmelt, and the other is older, more salty ground water. Second, we assume that stream water at a given point at any time is some mixture of these two water types. Last, we assume that ground water is added to the stream at a constant rate, whereas rainwater or snowmelt water is added on a variable, day-to-day basis. Essentially, then, stream water becomes a variable mixture between pure rainwater or snowmelt water and pure ground water, the proportions of which vary from day to day and season to season. During flood periods, say during the spring snowmelt, stream water is composed of nearly pure snowmelt water. During drought periods, say in midsummer, stream water is composed of nearly pure ground water. In spite of these tightly prescribed and perhaps simplistic assumptions, the model does in fact work and serves to reliably and accurately describe quantitative fluctuations in stream-water chemistry on a day to day or season to season basis (Johnson et al., 1969). Within an undisturbed watershed the model predicts quite precisely the changes in stream-water chemistry elicited by a given storm or a given dry period. The simple premises of the model testify to the remarkable chemical stability and chemical controls inherent in these forested ecosystems. It is significant in this regard that the model completely breaks down when the watershed is deforested (Likens et al., 1970).

The maintenance of relatively constant stream-water chemistry is in part a result of the year-long high infiltration capacity and permeability of the forest soils at the HBEF, and in part of a high rate constant for the chemical reactions manifested in the soil zone. Apparently, chemical equilibria between aqueous and solid phases are achieved rapidly through various geologic and biologic reactions in the soil. As a consequence, stream-water chemistry is essentially established in the soil zone or the interflow drainage passages. This is an excellent example of the way in which both the form and rate of chemical flux are modified and tightly regulated by a forest ecosystem. Similar constancy of stream-water chemistry has been observed for the major cations in the Coweeta Experimental Forest of North Carolina (Johnson and Swank, 1973).

These examples suggest that the relative independence of stream-water chemistry from stream-water discharge is a rather general feature of eastern deciduous forest ecosystems on granitic substrates. The importance of this should not be underestimated, for it demonstrates that the chemistry of regional headwater streams is the product of the natural ecosystem, the control of which may be altered by disturbance. For example, in disturbed forested areas and agricultural lands concentrations

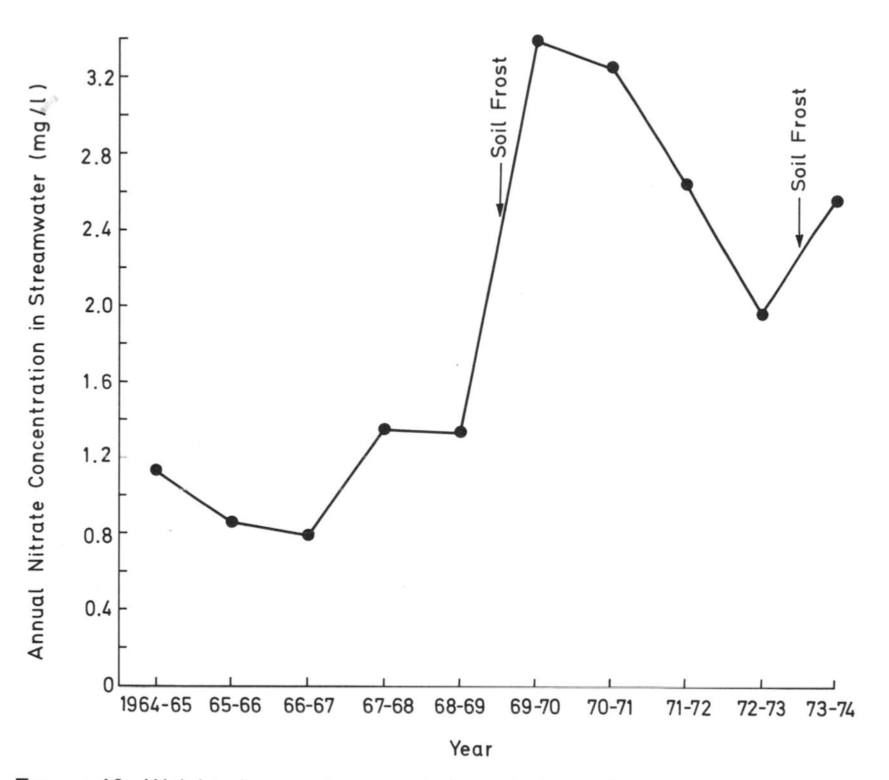

FIGURE 16. Weighted annual concentration of nitrate in stream water for undisturbed watersheds at Hubbard Brook.

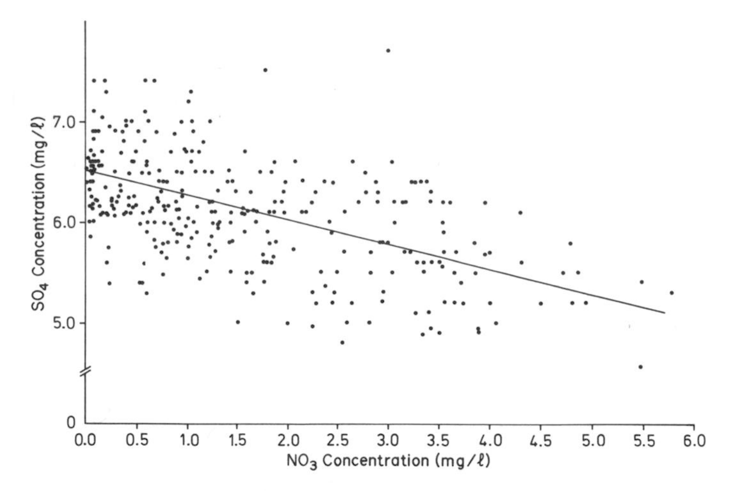

FIGURE 17. Relationship between nitrate and sulfate concentrations in stream water. Data were obtained weekly during October through May from 1964 to 1974. The *F*-ratio for this regression line is significant ($p < 0.001$ for a larger *F*-value) and the correlation coefficient is 0.54.

of chemicals in stream water usually increase and vary markedly with discharge (e.g., Likens and Bormann, 1974a; Likens et al., 1970; Powell, 1964, pp. 18–19).

Although the concentration of most ions in stream water varies little from year to year (note the standard error of annual arithmetic means, Table 8), the annual weighted concentration of nitrate was generally increasing during the period 1964–1974 (Figures 16 and 18). For example, the average weighted concentration during 1964–1967 was 0.92 mg/l; during 1967–1969 it was 1.40 mg/l, and in 1969–1972 it was 3.14 mg/l. An explanation for this change in the nitrate concentration is primarily circumstantial and there are several possibilities. The snow cover formed very late in the winter of 1969–1970 and for the first time on record widespread freezing of the soils was observed. Annual stream-water concentration of nitrate rose markedly that year (Figure 16). Then, in 1973–1974, widespread freezing of the forest soils occurred again, and again nitrate concentration increased. Various workers have reported that the freezing and thawing of soils promotes nitrification and the

FIGURE 18a and b. Monthly concentrations of nitrate in precipitation (a) and stream water (b) for Watershed 6 during the period 1964–1974. The regression line is significant ($p < 0.001$ for a larger *F*-value) and the correlation coefficient is 0.33.

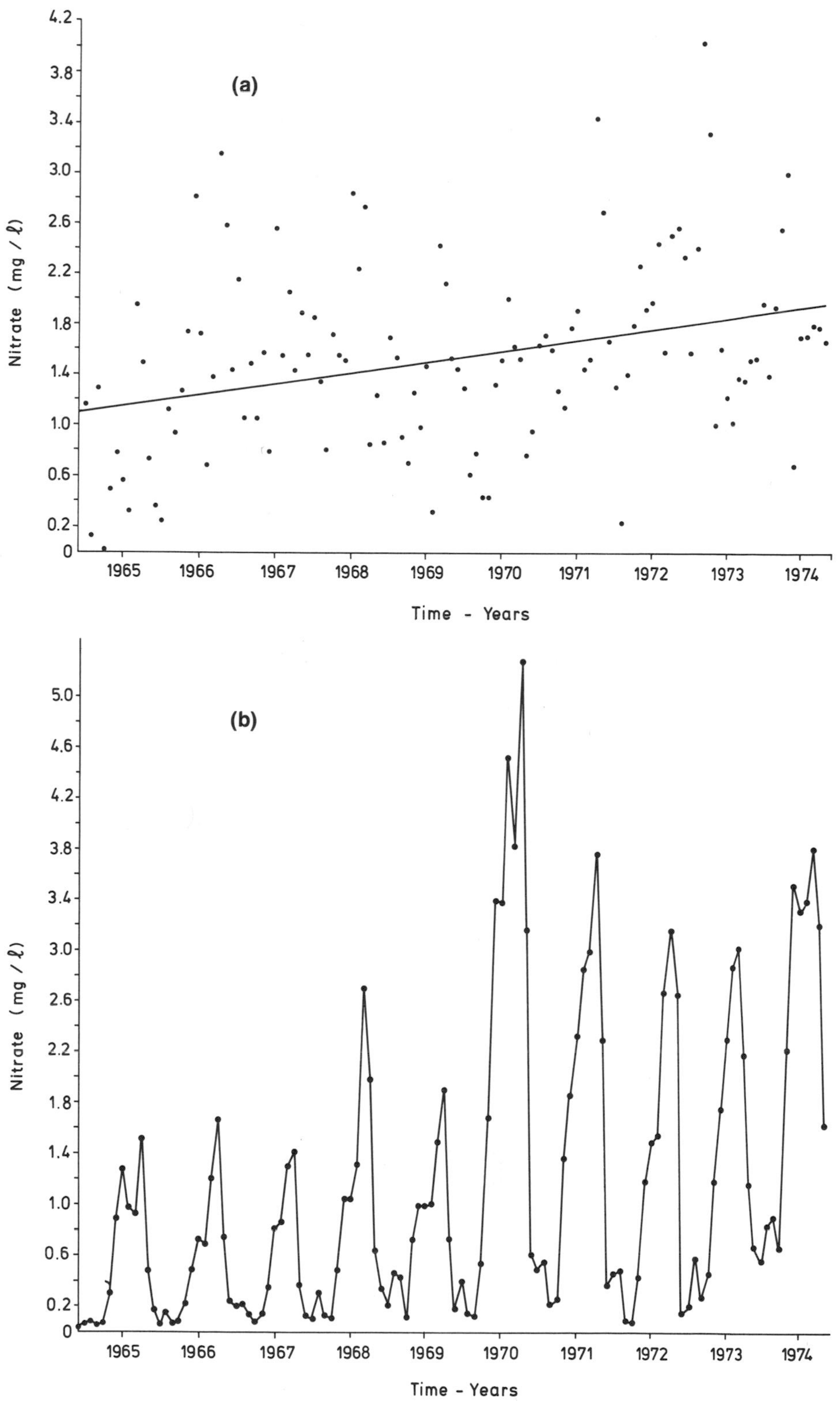

(a)
Nitrate (mg / ℓ)
4.2
3.8
3.4
3.0
2.6
2.2
1.8
1.4
1.0
0.6
0.2
0
1965
1966
1967
1968
1969
1970
1971
1972
1973
1974
Time - Years
(b)
Nitrate (mg / ℓ)
5.0
4.6
4.2
3.8
3.4
3.0
2.6
2.2
1.8
1.4
1.0
0.6
0.2
0
1965
1966
1967
1968
1969
1970
1971
1972
1973
1974
Time - Years

mobilization of nitrate to drainage waters (Arefýeva and Kolesnikof, 1964; references cited in McGarity and Rajaratnam, 1973). Therefore, freezing and thawing of the Hubbard Brook soils may have induced nitrification and the subsequent loss of relatively large amounts of nitrate in stream water. However, we have no explanation for the very slow rate of recovery (residual effect) observed during the 2 yr after the freezing occurred in 1969–1970 . It might have been expected that if soil frost enhanced nitrification, the nitrate would have been flushed rapidly from the system and in subsequent years nitrate concentrations in stream-water would have returned to "normal."

An inverse relationship between nitrate and sulfate concentration in stream water has been observed in undisturbed ecosystems (Figure 17; Likens et al., 1970). This inverse relationship has a strong seasonal feature. Stream-water concentrations of NO_3^- generally reach a maximum during the late winter, whereas concentrations of SO_4^{2-} at that time are at a minimum. Although there has been no consistent decline in stream-water sulfate concentrations since 1968, sulfate concentrations in 1969–1970 were about 10% less than the long-term average and the input of sulfate during 1968–1972 was generally lower than in other years (see Table 11). It may be that part of the increased nitrate concentration in stream water after 1968 is related in some manner to the sulfur biogeochemistry of the ecosystem. This deserves further study.

Another possible explanation for the increasing concentration of nitrate in stream water during the period of our study is that nitrate concentrations in precipitation have also been increasing (Figures 9 and 18). However, there is very poor correlation between the pattern for annual values of nitrate in precipitation (Figure 9) and in stream water (Figure 16); that is, nitrate concentrations in precipitation were relatively low in 1969–1970, whereas this was the peak year for stream-water concentrations.

A final complicating factor is that some watersheds show a larger response for nitrate concentrations in stream water than others. For example, the change in nitrate concentration in stream water from 1972–1973 to 1973–1974 was greatest in Watershed 1 (40% increase) and somewhat less in Watershed 6 (28% increase) and Watersheds 3 and 5 (25% increase). The most obvious difference between these watersheds is in area: Watershed 1, 11.8 ha; Watershed 6, 13.2 ha; Watershed 5, 21.9 ha; Watershed 3, 42.4 ha; even so, a causal relationship is not clear, particularly for the larger Watersheds, 3 and 5.

Calculation of Mass Output of Dissolved Substances

The output of dissolved substances from our watershed ecosystems is calculated in a manner analogous to calculations of input in bulk precipitation as described above. Stream-water outputs are expressed in mass units (kg) or ionic units (Eq) per hectare of watershed ecosystem.

Relationship of Mass Nutrient Output to Annual Streamflow

Since the chemical concentrations are relatively constant in stream water, the gross annual export of most ions is directly related to the annual streamflow and is highly predictable. The regressions relating gross output of calcium, magnesium, potassium, and sodium to annual discharge are all positive and highly significant (probability for a larger *F*-value < 0.001; Figure 19). The total equivalent export for the sum of all cations also can be predicted accurately from annual streamflow data (Figure 20). This is to be expected because the outputs of the major cations in stream water are significantly related to annual streamflow (Figure 19).

The high predictability of mass output in stream water represents one of the major findings of the Hubbard Brook Ecosystem Study. Because of the regulating effect of the forest ecosystem on chemical concentrations in

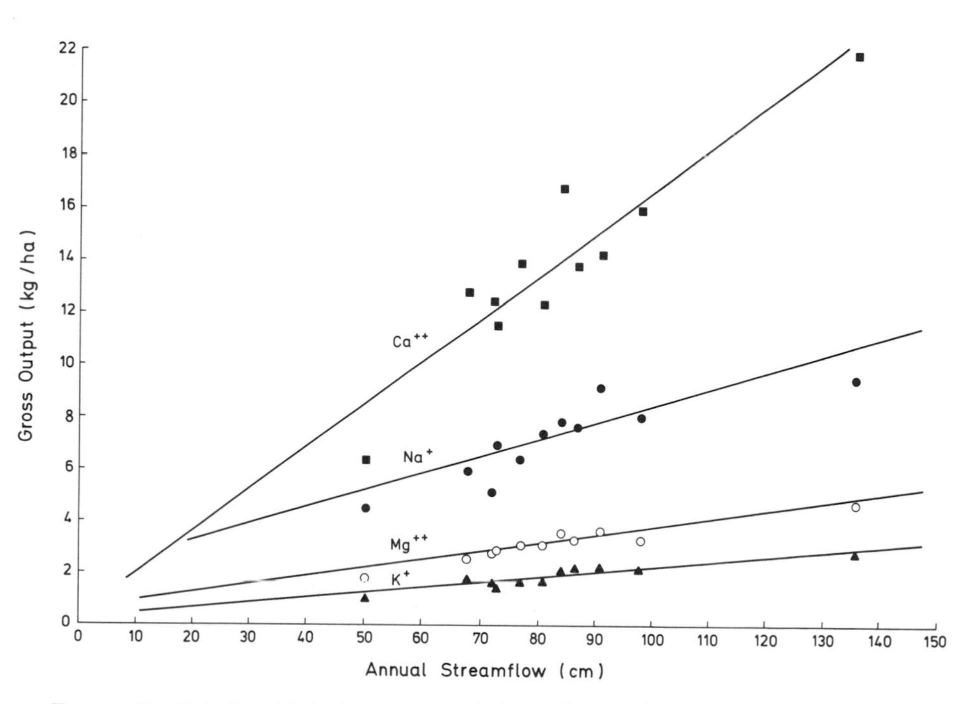

FIGURE 19. Relationship between annual streamflow and gross output of calcium, sodium, magnesium, and potassium (kg/ha) during 1963–1974 for undisturbed Watersheds 1–6 of the Hubbard Brook Experimental Forest. The regression lines fitted to these data are: $Y = b + ax$, where Y = gross output in kg/ha, b = Y intercept, a = slope, and x = annual streamflow in 10^6 l/ha. The equations for each line are: Ca^{2+}, $Y = 0.37 + 1.61\ x$ (*F*-ratio of 56.4^{***}; correlation coefficient, 0.93); Na^{+}, $Y = 1.92 + 0.64\ x$ (*F*-ratio of 39.8^{***}; correlation coefficient, 0.90); Mg^{2+}, $Y = 0.57 + 0.31\ x$ (*F*-ratio of 82.2^{***}; correlation coefficient 0.95); K^{+}, $Y = 0.24 + 0.196\ x$ (*F*-ratio of 52.0^{***}; correlation coefficient, 0.92).

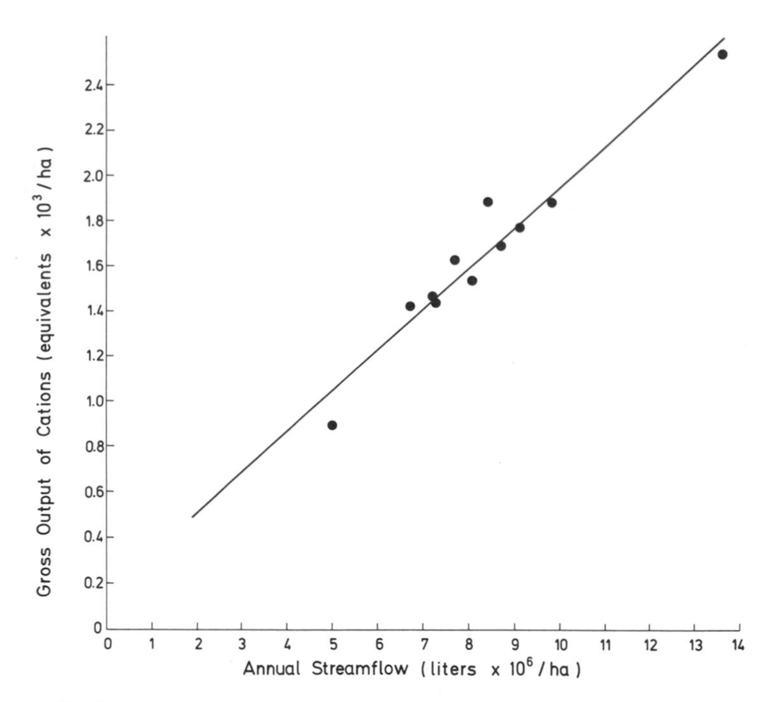

FIGURE 20. Relationship between annual streamflow and gross output of cations (10^3 Eq/ha) during 1963–1974 in undisturbed Watersheds 1–6 of the Hubbard Brook Experimental Forest. The regression line fitted to these data is $Y = b + ax$, where Y = gross output of cations in 10^3 Eq/ha, b = Y intercept, a = slope, and x = annual streamflow in 10^6 l/ha. The equation for the line is $Y = 0.15 + 0.18\,x$ (F-ratio of 142^{***}; correlation coefficient, 0.97).

stream water as it emerges from the system, mass output can be predicted simply from a knowledge of annual hydrologic output. Not only is this relationship important in characterizing conditions in an undisturbed forest and thus providing a baseline vital to measurement of any changes resulting from forest manipulation, but it also provides information useful to land-use and hydrologic planners.

Annual Variation in Mass Output of Dissolved Substances

On a mass basis (kg/ha) sulfate and dissolved silica are the dominant substances exported in stream water, whereas on an ionic basis sulfate and calcium dominate. The annual gross output of total dissolved substance (excluding dissolved organic carbon) during 1963–1974 averaged 147.2 kg/ha; excluding dissolved silica, it was 109.6 kg/ha (Table 11). The predominance of sulfate is striking. On the average, the annual gross output of sulfate is 52.8 kg/ha or 1.1×10^3 Eq/ha, which represents, on the average, 36% of the total dissolved substances or 68% of the anionic

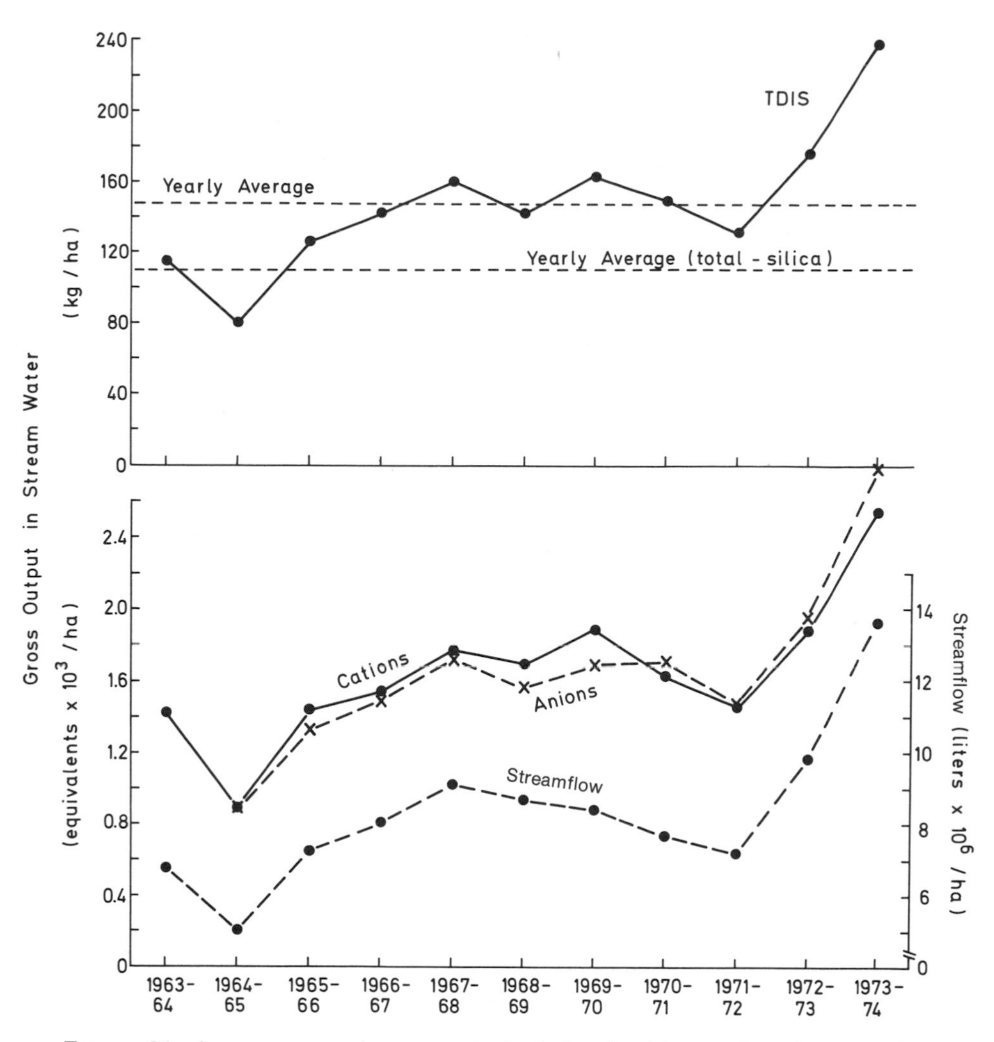

FIGURE 21. Average annual gross output of dissolved inorganic substances in stream water from undisturbed Watersheds 1–6 of the Hubbard Brook Experimental Forest. TDIS = total dissolved inorganic substances.

equivalency. The annual gross output of total inorganic dissolved substances varies relatively little from year to year (Figure 21). Annual mass export varied by threefold and equivalency by 2.9-fold, which could be largely explained by the range (2.7-fold) in annual streamflow (Figure 4). The largest output of ions occurred during the 1973–1974 water-year (60.4% greater than the long-term average) and the drought year of 1964–1965 had the smallest output (45.4% less than the long-term average; Figure 21). The average annual output of cations during the period 1963–1964 to 1973–1974 was calculated to be 1.65×10^3 Eq/ha, and the average annual output of anions during 1963–1974 was 1.61×10^3 Eq/ha. The poorest agreement in equivalents, 11% (difference based on the smaller value), between cations and anions (Figure 21) occurred in 1969–1970 and the best agreement, 1%, was in 1964–1965 and 1971–1972.

Sizable amounts of dissolved organic carbon are lost from the undisturbed Hubbard Brook watersheds each year in stream water. Based on an average carbon concentration of 1 mg/l (Hobbie and Likens, 1973), about 8 kg/ha of dissolved organic carbon are exported each year.

Particulate Matter

Measurement of Mass Output

In addition to dissolved chemical substances, both organic and inorganic particulate matter may be exported from ecosystems in stream water. Particulate matter is removed from the watershed as suspended load carried by turbulent water and as bedload rolled, slid, or bounced along the stream bed. Heavier particulate matter collects in the ponding or stilling basin behind the weir where, over an 8-yr period, it was periodically dug out, weighted, proportionally sampled, oven dried, and analyzed for dry weight and organic and inorganic content (Bormann et al., 1969, 1974). Suspended particulate matter that passed over the weir was periodically sampled using a 1-mm mesh net and by passing a sample of water through the net and then filtering it through a 0.45 μm Millipore filter at 40 psi. Samples were then analyzed for dry weight and organic and inorganic content and expressed as concentrations per unit volume of water. These concentrations were used in combination with the hydrologic discharge record to determine the export of materials over the weir. These three components, basin, netted, and filtered, were combined to give total particulate export from the ecosystem (Figure 22).

It is noteworthy that this estimate of total export includes both suspended particulate matter and bedload, whereas most reports of particulate export from ecosystems are concerned with suspended load alone. If the heavier inorganic material that collects behind the weir is considered as a measure of bedload, then bedload constitutes about 55% of the total export.

Annual Loss

Erosion and transport of particulate matter from these forested watersheds is relatively low even though the watersheds are on steep slopes (12–13°) and are subject to large amounts of precipitation (130 cm/yr). Although the glacial till in these Watersheds is relatively resistant to erosion (Hunt, 1967) it is primarily the living and dead biomass that minimizes the loss of particulate matter by regulating the amount, timing, and effect of moving

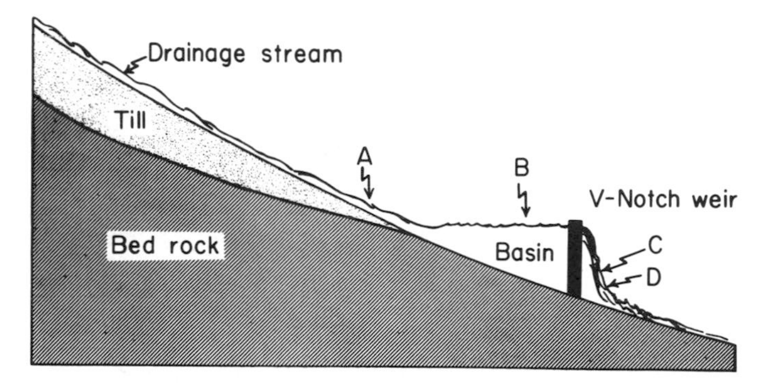

FIGURE 22. Sampling procedure employed in the measurement of total nutrient losses from a watershed ecosystem. *A* = water sample for dissolved substances, *B* = suspended and bedload dropped in basin, *C* = water sample passed through net for millipore filtration, *D* = net sample. Total losses = dissolved substances (*A*) + particulate matter (*B* + *C* + *D*). After Bormann et al. (1969).

water within the undisturbed ecosystem. Organic debris dams in the stream channels of these headwater ecosystems apparently play a major role in minimizing the export of particulate matter.

The undisturbed ecosystems tightly regulate erosion and transport and lose only about 33 ± 13.4 kg/ha of particulate matter with annual runoff (Table 9). This material averages about 33 ± 4% (standard error) organic, but the organic:inorganic proportion is strongly influenced by flow rate of the stream, with higher inorganic proportions associated with higher flow rates (Figure 23; Bormann et al., 1974).

Seasonal Variation in Erodability

Although the maximum output of particulate matter occurs during high-discharge periods, summer flows are more effective in exporting particulate matter than are flows of about the same velocity when vegetation is dormant (Bormann et al., 1974). The 30% greater erodability during the summer is probably related to increased biologic activity and decomposition within the ecosystem. In winter the stream channels are often stabilized by ice and snow and, in addition, precipitation generally falls as snow, with low potential energy of impact.

Particulate Matter Versus Dissolved Substance Export

There is a sharp contrast in the response of dissolved substance and particulate matter export to discharge rate of the stream. Concentrations of dissolved substances are relatively little affected by flow rates, whereas particulate matter concentrations are directly and exponentially related to stream discharge (Figure 24). As a consequence, the bulk of particulate

TABLE 9. Annual Particulate Matter Output in Kilograms of Oven-Dry Weight Organic and Inorganic Materials per Hectare for Watershed 6[a].

	Source of output	Organic	Inorganic	Total
1965–1966	Ponding basin	2.12	1.77	3.89
	Net	0.34	0.01	0.35
	Filter	1.37	1.28	2.65
	Total	3.83	3.06	6.89
1966–1967	Ponding basin	13.41	17.07	30.48
	Net	0.39	0.01	0.40
	Filter	2.72	2.95	5.67
	Total	16.52	20.03	36.55
1967–1968	Ponding basin	3.83	5.93	9.76
	Net	0.43	0.01	0.44
	Filter	2.61	2.82	5.43
	Total	6.87	8.76	15.63
1968–1969	Ponding basin	4.61	8.31	12.92
	Net	0.42	0.01	0.43
	Filter	2.57	2.81	5.38
	Total	7.60	11.13	18.73
1969–1970	Ponding basin	11.28	30.67	41.90
	Net	0.40	0.01	0.41
	Filter	3.30	3.69	6.99
	Total	14.98	34.37	49.30
1970–1971	Ponding basin	2.70	2.75	5.45
	Net	0.39	0.01	0.40
	Filter	1.77	1.78	3.55
	Total	4.86	4.54	9.40
1971–1972	Ponding basin	3.66	2.52	6.18
	Net	0.37	0.01	0.38
	Filter	1.76	1.80	3.56
	Total	5.79	4.33	10.12
1972–1973	Ponding basin	17.88[b]	77.40	95.28
	Net	0.60	0.01	0.61
	Filter	10.09	13.61	23.70
	Total	28.57	91.02	119.59
8-yr average	Ponding basin	7.44	18.30	25.74
	Net	0.42	0.01	0.43
	Filter	3.27	3.84	7.11
	TOTAL	11.13	22.15	33.28

[a]Ponding basin output = particulate matter collected in a ponding basin of the weir upstream from the V-notch; Net output = suspended particulate, >1mm, that passes over the V-notch of the weir; Filter output = suspended particulate matter, >0.45μm <1mm, that passes over the V-notch of the weir.

[b]Underestimate.

matter is moved during storms. Output of dissolved substances is therefore closely related to annual output of water, whereas removal of particulate matter is more of a stochastic process, strongly related to the occurrence of random storms. This is reflected in the fact that during the period

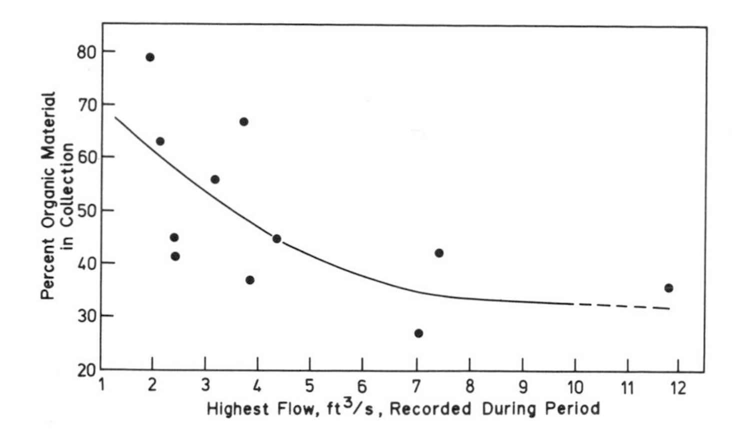

FIGURE 23. Percent organic material in basin collections from Watershed 6 plotted against the highest flow recorded during the period. One ft^3/sec equals 28.3 l/sec. After Bormann et al. (1974).

1965–1973 particulate matter output ranged from 7 to 120 kg/ha·yr and was highly correlated with the occurrence of individual storms in a particular year. An analysis of more than 4 yr of data indicated that 86% of the total particulate matter was exported in 1.6% of the total time in the period and with 23% of the water; 16% was exported in 0.0025% of the time period with 0.2% of the water (Bormann et al., 1974).

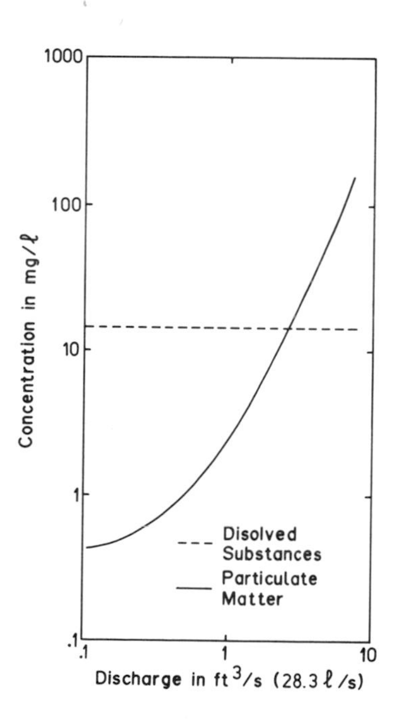

FIGURE 24. General relationships between the concentration of dissolved substances and particulate matter and stream discharge in the Hubbard Brook Experimental Forest. After Bormann et al. (1969).

TABLE 10. Average Annual Gross Output of Some Elements as Dissolved (*D*) and Particulate (*P*) Substances from Watershed 6 of the Hubbard Brook Experimental Forest[a].

Element	Particulate + dissolved element total, kg/ha	Particulate		Dissolved	
		kg/ha	% of *P* + *D* element total	kg/ha	% of *P* + *D* element total
Aluminum	3.37	1.38	40.9	1.99	59.1
Calcium	13.93	0.21	1.7	13.7	98.3
Chloride	4.58	—	0	4.58	100
Iron	0.64	0.64	100	—	0
Magnesium	3.34	0.19	5.7	3.15	94.3
Nitrogen	4.01	0.11	2.7	3.90	97.3
Phosphorus	0.019	0.012	63.2	0.007	36.8
Potassium	2.40	0.52	21.7	1.88	78.3
Silicon	23.8	6.19	26.0	17.6[b]	74.0
Sodium	7.48	0.25	3.3	7.23	96.7
Sulfur	17.63	0.03	0.2	17.6	99.8
Carbon[c]	12.3	3.98[d]	32.4	8.35[e]	67.5

[a]Particulate matter losses during 1966–1967 to 1969–1970 are modified from Bormann et al. (1974). Dissolved substance losses are averages during the period 1963–1974.

[b]Assuming that dissolved silica is in the form of SiO_2.

[c]Organic only.

[d]Assuming organic matter is 40% C.

[e]Assuming an average carbon concentration of 1 mg/l (Hobbie and Likens, 1973).

These relationships have important implications for the export of certain elements. Most of the chemical elements with a sedimentary cycle are exported from the watershed ecosystem primarily in the dissolved form, as shown above; however, the majority of the iron and phosphorus is lost in the form of particulate matter (Table 10). In contrast to other elements, therefore, the output of iron and phosphorus is more directly related to stream discharge rate and is less predictable from annual streamflow. Moreover, phosphorus in particulate form is unavailable to aquatic microorganisms and plants within the stream ecosystem unless it first undergoes decomposition or mineralization. This linkage and the relationships between the losses of particulate and dissolved phosphorus in stream water are currently under study in the HBEF.

Our data show that in the forest ecosystem at Hubbard Brook, solution losses are the major force in the geologic process of fluvial denudation. In terms of gross export of mass, dissolved substances—about 165 kg/ha, including dissolved organic matter—are about five times greater than particulate matter losses (Tables 9 and 11).

The Role of Debris Avalanches in Landscape Denudation

Our data suggest that headwater watersheds, with their well-developed forests, are gradually lowered in place by the action of solution coupled with slow mass movements. Both of these actions deliver material to the stream, where it is removed by erosion and transportation. The possibility that infrequent debris avalanches play an equal or greater role in moving material downslope in headwater watersheds should also be considered. For example, Hack and Goodlett (1960) have shown in the Appalachian Mountains of Virginia and West Virginia that infrequent but catastrophic landslides play a major role in removing material from low-order watersheds and that this activity is independent of any restraining influence by the biota. Flaccus (1958a,b) has shown essentially the same thing for the White Mountains of New Hampshire. Our data on stream losses allow a rough comparison of these very different denudational processes.

Flaccus (1958a,b) mapped 543 debris avalanches on aerial photos of the White Mountains. His study area, most of which is heavily forested, is adjacent to the Hubbard Brook Valley and contains extensive stretches of the Littleton formation (now mapped as the Rangeley formation in our watershed). The mapped avalanches occur in about equal proportions on slopes of the Littleton formation and slopes underlain by plutonic rocks. Flaccus examined in detail a range of avalanches, thought to be a fair sample of the total, and calculated the average weight of material moved downslope as 21,800 metric tons. In those slides fresh enough to permit location of the topmost elevation it was found that 100% occurred above the 610 m elevation contour. Within his study we have calculated an area of about 128,000 ha above the 610 m elevation. Using Flaccus' data for numbers and dates of avalanches and average amounts of material moved in each event, we have estimated that 462 kg/ha·yr are moved downslope by debris avalanches occurring above 610 m in the White Mountains (Bormann et al., 1969). This estimate is about triple our estimate of 200 kg/ha·yr for dissolved plus particulate export from headwater watersheds of the HBEF. This suggests that debris avalanches are at least equivalent to the combined action of solution and slower mass movements in lowering the surface of low-order watersheds in the White Mountains.

However, two additional factors must be considered. (1) Disturbances that destroy the biotic stability of the ecosystem, such as fire, can lead to greatly increased losses by solution, mass wasting, and erosion; and (2) the incidence of avalanches per unit area increases sharply with increasing elevation and increasing slope (Flaccus, 1958a,b). In the White Mountains, only 18% of the recorded avalanches occur wholly below 910 m.

At Hubbard Brook it seems unlikely that debris avalanches have occurred within recent times in the area currently occupied by northern hardwood forest. With minor exceptions, elevations within the Hubbard Brook Valley are below 910 m and except for a few sites of very restricted area our slope does not meet Flaccus' minimum slope requirement of 25°, nor is there any evidence of recent or old debris avalanches. It seems safe to conclude that debris avalanches have not played a significant role in the denudation of the relatively gentle slopes of the Hubbard Brook watershed, at least during the last millenium.

Apparently, denudation in the Hubbard Brook area is primarily a result of the less dramatic action of solution and erosion in combination with slower mass movements, such as creep, which may deliver materials to the streambeds. This, in turn, suggests that the weathering rind in our watersheds is fairly old. Locally, this is shown by well-developed soil profiles; however, in some places the surface is subject to shallow stirring by burrowing animals and from trees uprooted by the wind.

Addendum

Snow and Snowpack Chemistry

The chemistry of the considerable amount of snow that frequently accumulates as snowpack during the winter at Hubbard Brook may have little similarity to the chemistry of the snow that falls (Hornbeck and Likens, 1974), but can significantly affect the chemistry of stream water during snowmelt (e.g., Johnson et al., 1981; Hooper and Shoemaker, 1985; Hooper, 1986). For example, potassium concentrations tend to be up to two orders of magnitude higher in the snowpack than in ambient snowfall. In addition to concentration due to sublimation, much of this increase comes from the leaching of potassium from plant parts also incorporated in the snowpack (Hornbeck and Likens, 1974). In sharp contrast, amounts of nitrate and ammonium in the snowpack tend to be lower than amounts input in ambient snowfall.

Dry Deposition

In 1977 we were aware of the dry deposition of particles and gases from the atmosphere, but had little quantitative information relative to such fluxes for the ecosystems at Hubbard Brook (e.g., Eaton et al., 1976). Because dry deposition represents a potentially important input of nutrients and pollutants to forest ecosystems, much effort has been expended to quantify this flux at Hubbard Brook. Using the long-term record and a mass balance approach, Likens et al. (1990) estimated that for sulfur, some 37% of the total (wet plus dry) deposition annually occurred as dry deposition. Annual dry deposition varied from 12% of the total, in 1964–1965 to 61% in 1983–1984, and the average annual flux was estimated as about 420 eq SO_4^{2-}/ha·yr (Likens et al., 1990).

Lovett et al. (1992) have compared the results of three independent methods for estimating dry deposition of sulfur at Hubbard Brook: the mass balance approach, a throughfall approach, and an inferential method using measured atmospheric concentrations and assumed deposition velocities. The results of these approaches give annual values that differ by about fourfold, with the mass balance approach producing the largest value (Lovett et al., 1992). Each of these approaches has assumptions and errors associated with it, and dry deposition remains an extremely complicated flux to measure quantitatively in forest ecosystems. Currently, scientists at Hubbard Brook operate two air quality monitoring sites (at 250 m and 660 m elevation). One of these sites is affiliated with the National Clean Air Status and Trends Network (CASTNET) and uses the inferential method for estimating dry deposition.

Long-Term Trends

The longer-term perspective of biogeochemical records now spanning some 31 years has revealed some remarkable trends in the chemistry of atmospheric deposition and stream water at Hubbard Brook. Notably, annual, volume-weighted concentrations of hydrogen ion, sulfate, and calcium in bulk precipitation decreased by 40%, 37%, and 75%, respectively, since 1963 or 1964 (Figures 25 and 26). Since there was no significant temporal trend in the amount of annual precipitation (Figure 25), the annual inputs of these ions also decreased significantly during this period (Likens, 1992). The decline in base cations ($C_B = Ca^{2+} + Mg^{2+} + Na^+ + K^+$) in precipitation recorded at Hubbard Brook since 1963–1964 has also been found more widely in the eastern U.S. and in north-western Europe, although the length of record at these other locations is shorter (Hedin et al., 1994). Most of the decline in C_B at Hubbard Brook was due to a decline in Ca^{2+} concentrations and most occurred prior to about 1975 (Figure 26). By 1974 it appeared that the annual nitrate concentration in bulk precipitation was on an upward trend (Figure 9), but the longer-term record indicates that since about 1972 nitrate concentrations have been relatively flat (Figure 25).

The long-term record for concentration of hydrogen ion in precipitation is particularly interesting (Figure 27). Clearly the average precipitation at Hubbard Brook is not as acid in the 1990s as it was in the 1960s, but it required 18 years of record to fit a statistically significant linear regression ($p < 0.05$) to these temporal data (Likens, 1989). A close examination of this long-term record, however, shows periods comprising several years of data with different trends (Figure 27). Obviously, short-term (3–5 yr) records can be misleading relative to the overall trend. This is an important finding not only for interpreting results from the HBES, but also for general ecological studies and state and national monitoring programs.

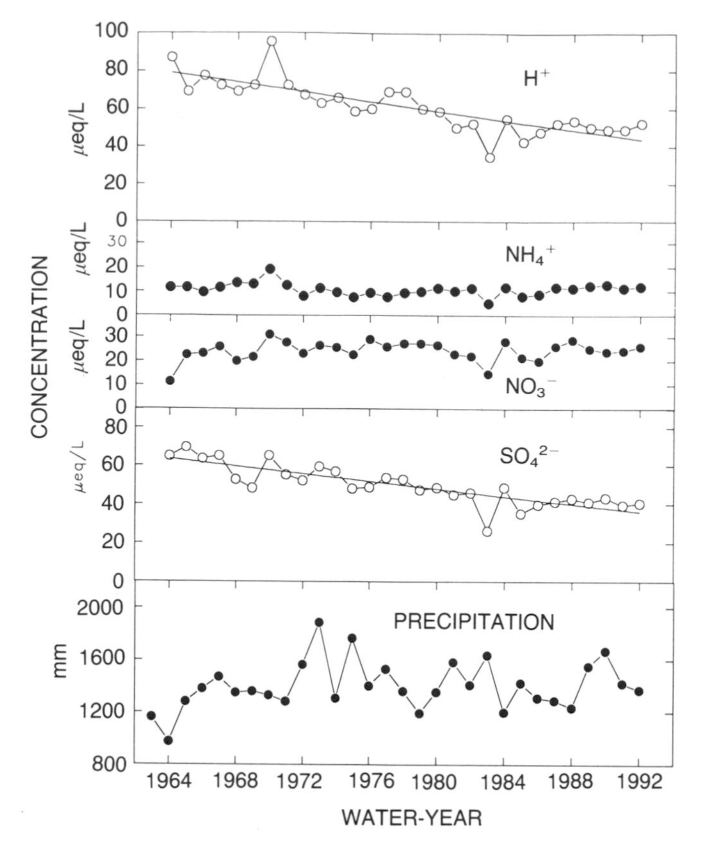

FIGURE 25. Annual volume-weighted concentration of H^+, NH_4^+, NO_3^-, and SO_4^{2-} in precipitation and amount of precipitation for Watershed 6 of the Hubbard Brook Experimental Forest between 1963–1964 and 1992–1993. The linear regression lines have a probability for a larger F-value of <0.001 and correlation coefficients of H^+, 0.66; SO_4^{2-}, 0.69.

Coincident with the declines in annual, volume-weighted concentrations in bulk precipitation were declines in the chemistry of stream water at Hubbard Brook. Average Ca^{2+} and SO_4^{2-} concentrations declined by 48% and 22% respectively during 1963 or 1964 to 1993 (Figures 28 and 29). Calcium concentrations in stream water actually increased from 1963–1964 to 1969–1970 and then decreased steadily thereafter. The decline in annual C_B and SO_4^{2-} concentrations in stream water was closely correlated with the decline in atmospheric inputs (e.g., Driscoll et al., 1989; Likens, 1992). The change in annual concentration of hydrogen ion in stream water was relatively small in the long-term record (Figure 29), primarily because the decline in base cations was paralleled by the decline in sulfate (Driscoll et al., 1989). In contrast, hydrogen ion

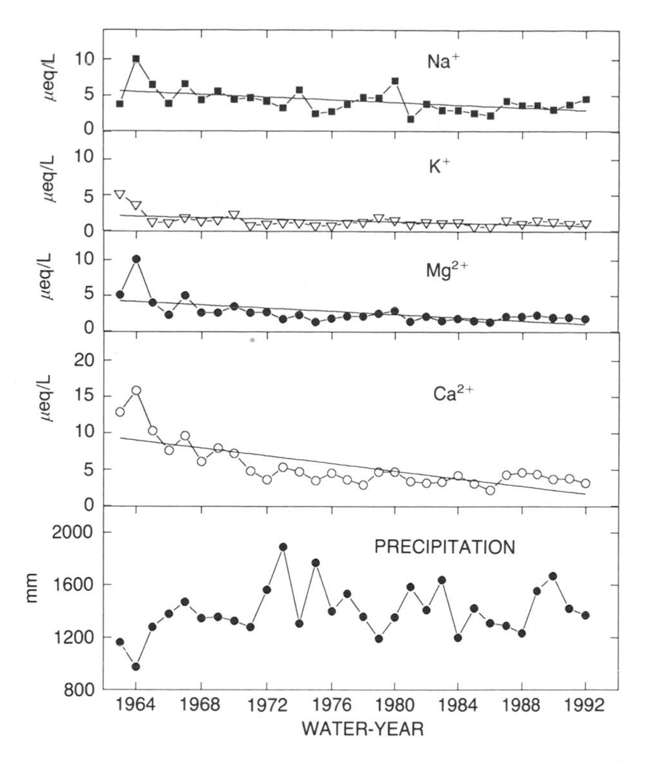

FIGURE 26. Long-term changes in concentrations of Na^+, K^+, Mg^{2+}, and Ca^{2+} in precipitation and in amount of precipitation for Watershed 6 of the Hubbard Brook Experimental Forest between 1963–1964 and 1992–1993. The probability for a larger *F*-ratio for all linear regression lines is <0.01, and correlation coefficients of: Na^+, 0.23; K^+, 0.21; Mg^{2+}, 0.34; Ca^{2+}, 0.54.

and nitrate concentrations in stream water and bulk precipitation were not closely correlated (Figures 25 and 29). Annual inputs of dissolved inorganic nitrogen (DIN) were more closely correlated to annual outputs of DIN in stream water (Figure 30). The explanation for the marked change in pattern in stream-water outputs of DIN during the long-term record (Figure 30) currently is the subject of much speculation and research at Hubbard Brook.

It is sometimes suggested that a few stream-water samples be collected to serve as a biogeochemical baseline for a terrestrial ecosystem. The marked and often unexplainable changes in the intensively sampled stream water at Hubbard Brook suggest caution in attributing baseline characteristics to a few samples that unwittingly may characterize a high or low period in the history of the site. Clearly, the most useful biogeochemical baseline is one of sufficient length to allow trend analyses.

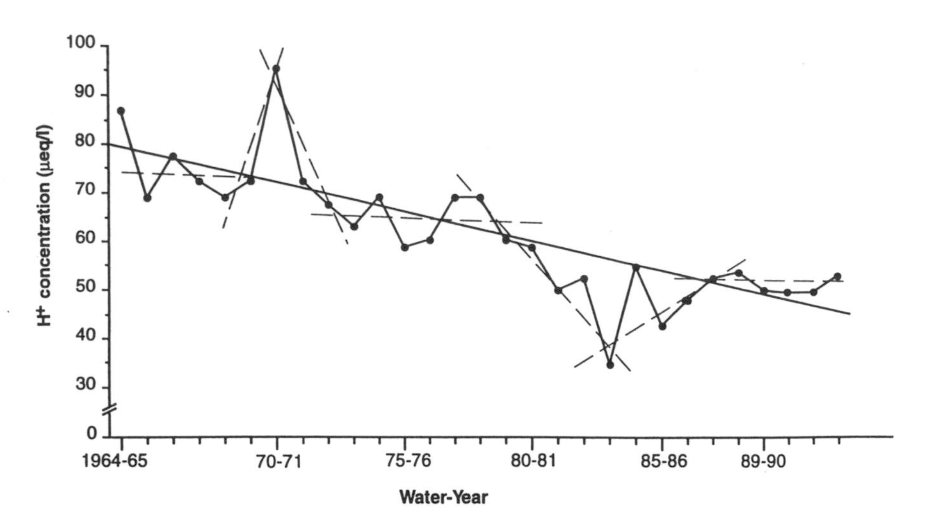

FIGURE 27. Annual, volume-weighted concentration (µeq/liter) of hydrogen ion for Watershed 6 of the Hubbard Brook Experimental Forest from 1964–1965 to 1992–1993. The linear regression line has a probability for a larger *F*-value of <0.001 and a correlation coefficient of 0.66. The shorter, dashed lines are fitted by eye. Updated from Likens, 1989.

Acid Rain

In 1977 the acid rain debate in the United States and Europe was in full swing. The First International Symposium on Acid Precipitation and the Forest Ecosystem had been held in 1975 at Ohio State University. Numerous studies of the ecological impact of acid rain were underway at Hubbard Brook, but the backbone of our research on this important environmental problem was the developing long-term record of precipitation and stream-water chemistry. Partly on the basis of the long-term record of precipitation chemistry at Hubbard Brook, the U.S. Congress enacted the amendments to the 1970 Clean Air Act (1990), which for the first time explicitly included provisions for regulating acid rain. The 1990 Amendments to the 1970 Clean Air Act call for a reduction of 10 million tons (9.1 million metric tons) in SO_2 emissions per year below the 1980 levels by the year 2000 (Figure 31). Prior to the Amendment, it was projected that emissions of SO_2 would increase to the year 2000 and somewhat beyond (Streets and Veselka, 1987). Thus, the *effective* reduction caused by the 1990 Amendments to the Clean Air Act is appreciably more than 10 million tons (9.1 million metric tons) per year.

Using long-term data from Hubbard Brook and the statistically significant ($p < 0.01$) relationship between emissions of SO_2 in the eastern U.S. (Figure 32) and SO_4^{2-} deposition at Hubbard Brook ($Y = -308 + 51.1\ X$ ($r^2 = 0.70$), where $Y = SO_4^{2-}$ deposition in eq/ha·yr and $X =$

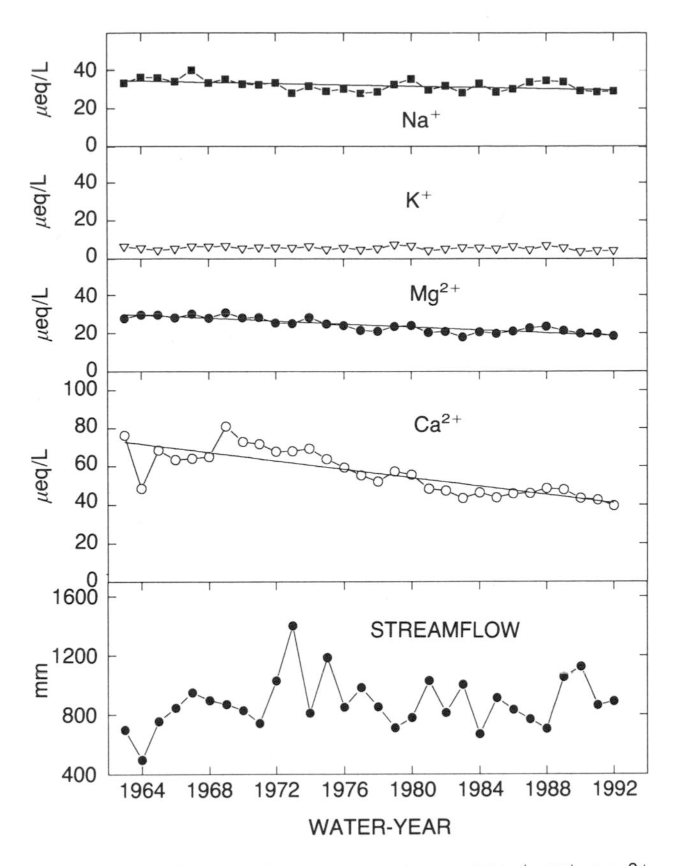

FIGURE 28. Long-term changes in concentrations of Na^+, K^+, Mg^{2+}, and Ca^{2+} in stream water and in amount of stream water for Watershed 6 of the Hubbard Brook Experimental Forest between 1963–1964 and 1992–1993. The probability for a larger *F*-ratio for all linear regression lines is <0.006, and correlation coefficients of: Na^+, 0.24; Mg^{2+}, 0.77; Ca^{2+}, 0.68.

SO_2 emissions in metric tons (yr^{-1}). This relation of sulfate deposition to emissions of SO_2 (Figure 33) can be used to extrapolate the long-term record of SO_4^{2-} deposition at Hubbard Brook to the year 2000 (Figure 34). These extrapolations represent wet deposition only and an estimate of dry deposition, equivalent to 50% of the long-term average (Likens et al., 1990) dry deposition value at Hubbard Brook, must be added at year 2000 (Figure 34).

Considering both of these scenarios for projecting wet and dry deposition of sulfate at Hubbard Brook to the year 2000, the estimated sulfur loading from the atmosphere will still be about three times higher than values recommended for protection of sensitive forest and associated aquatic ecosystems like those found at Hubbard Brook (see Likens,

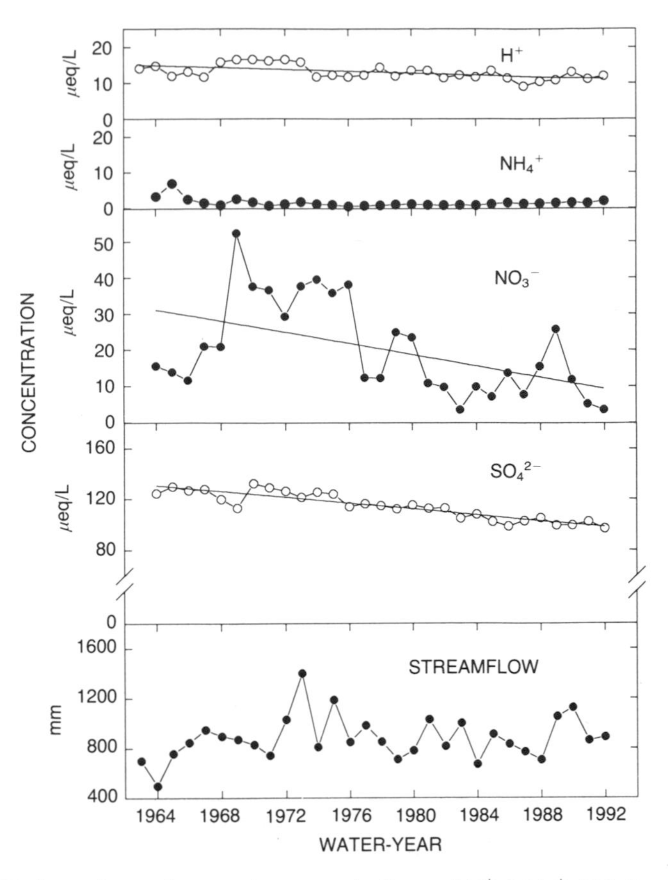

FIGURE 29. Long-term changes in concentrations of H^+, NH_4^+, NO_3^-, and SO_4^{2-} in stream water and in amount of stream water for Watershed 6 of the Hubbard Brook Experimental Forest between 1963–1964 and 1992–1993. The probability of a larger *F*-ratio for all linear regression lines is <0.005, and correlation coefficients of: H^+, 0.33; NO_3^-, 0.26; SO_4^{2-}, 0.83.

1992). Moreover, declining atmospheric inputs of base cations at Hubbard Brook (Figure 25; Driscoll et al., 1989; Hedin et al., 1994) apparently are causing forest and associated aquatic ecosystems at Hubbard Brook to become even more sensitive to atmospheric inputs of acidic substances (Driscoll et al., 1989).

How Long-Term Trends Can Define Research: An Example

A dramatic decline in forest growth during the last decade has been discovered at Hubbard Brook. Coincidentally, there has been a large

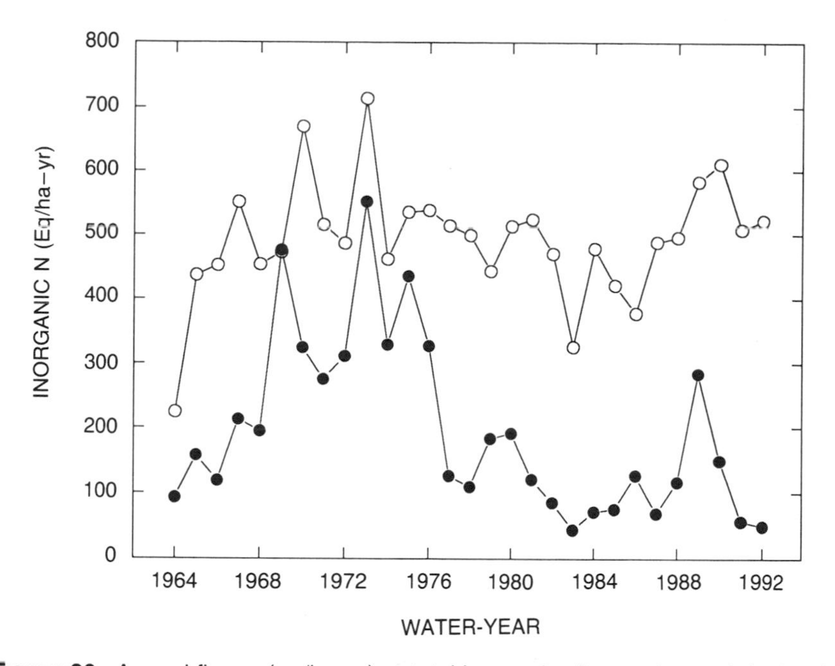

FIGURE 30. Annual fluxes (eq/ha·yr) or total inorganic nitrogen in precipitation (o) and stream water (•) in Watershed 6 from 1964–1965 to 1992–1993.

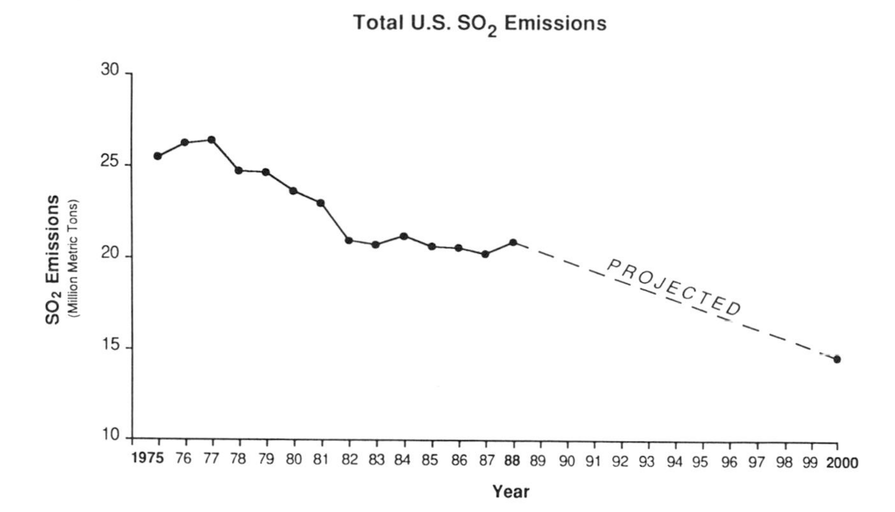

FIGURE 31. Total SO_2 emissions for the United States from 1975 to 1988 and projected to the year 2000, based on 1990 Amendments to the Clean Air Act. From Likens, 1992.

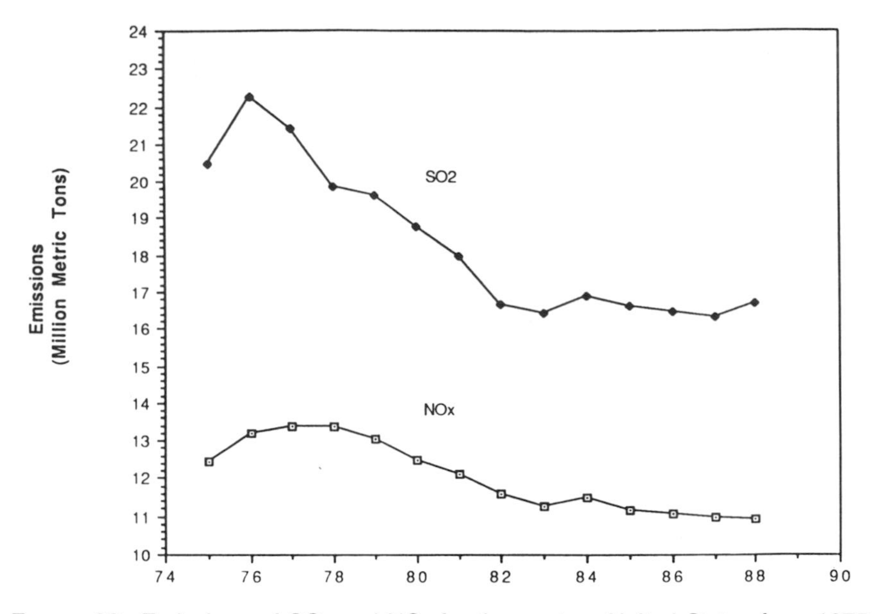

FIGURE 32. Emissions of SO_2 and NO_x for the eastern United States from 1975 to 1988. The area contributing these emissions is assumed to be the appropriate source area for the Hubbard Brook Experimental Forest. From Likens, 1992.

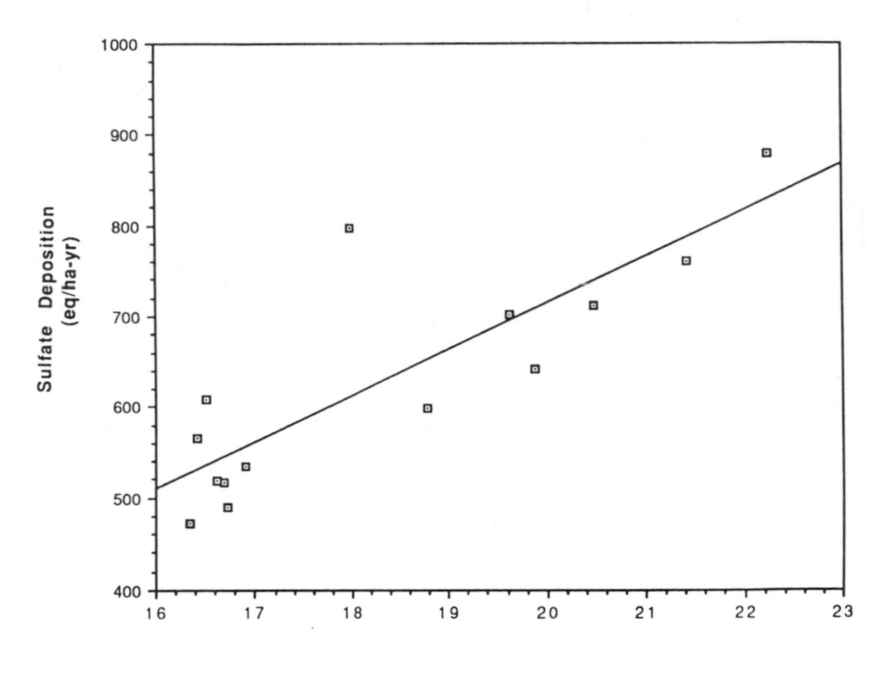

FIGURE 33. Relation between calendar-year SO_2 emissions in the eastern United States (Figure 32) and calendar-year SO_4^{2-} wet deposition at Hubbard Brook. The regression line has a probability for a larger *F*-value of <0.001 and an r^2 of 0.70. From Likens, 1992.

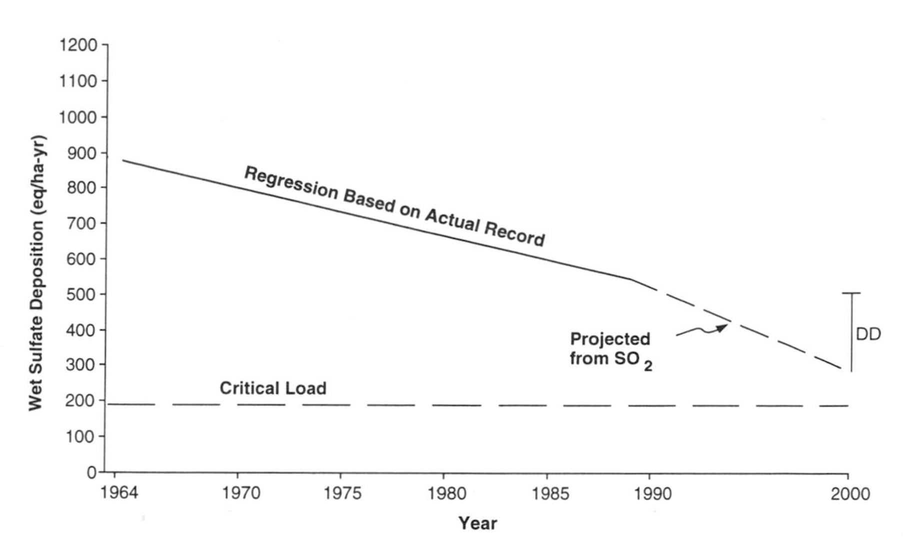

FIGURE 34. Wet deposition of sulfate for Watershed 6 of the Hubbard Brook Experimental Forest between 1964–1965 and 1989–1990 based on regression analysis (____). Projections to the year 2000 are based on regression relationship for historical changes in SO_2 emissions from 1975–1988 and wet SO_4^{2-} deposition (Figure 33) (__ __ __). DD = dry deposition of sulfate, Likens et al., 1990. (The critical load value is based on Nilsson and Grennfelt, 1988). Modified from Likens, 1992.

decline in calcium, a critical nutrient for the growth of trees, in various components of the forest ecosystem, including the forest floor. It also is particularly interesting that there has been a decline in the bird population in the Hubbard Brook area since about 1970 (R.T. Holmes, personal communication). These important trends have become evident through ongoing ecosystem research at Hubbard Brook.

The reasons for this major shift in ecosystem behavior are not yet evident, but ongoing research suggests that several factors may be interacting to limit forest growth. These are, natural factors (such as climatic and soil limitations on biomass accumulation), disease (such as beech bark disease), nutrient limitation (such as calcium or nitrogen availability), and air pollution effects (such as acid rain as a component of calcium deficiency). Indeed, all of these factors may be interacting in complex and enhancing ways to produce the decline in forest growth.

The emergent hypothesis is that during recent ecosystem development, a deficiency in available calcium in the soil has occurred, possibly as a result of the effects of acid rain. Therefore, we have proposed a comprehensive research effort to evaluate this hypothesis. Finding an answer to this critical question has great urgency because stagnating growth of forests is important to many segments of society, such as the forest industry, wildlife management, recreation, and the integrity of the ecosystem.

These studies would include long-term monitoring of precipitation, stream water, and soil chemistry, biomass accumulation, experimental plot studies at several size scales, an entire watershed manipulation, and modeling. To obtain maximum information useful to policy makers and those responsible for the management of the northern hardwood forest, this effort must be integrated within an ecosystem framework.

In our opinion a program mounted in this way would be the highest socially responsible investment of research funds for helping to resolve a potentially critical environmental problem.

4
Input–Output Budgets

Because our ecosystems are watersheds, the flux of chemicals into the system may be simplified to that in precipitation (meteorologic vector) and the output to that in drainage waters (geologic vector), at least for those nutrients without a prominent gaseous phase (Bormann and Likens, 1967; Likens and Bormann, 1972). Theoretically, then, the difference between annual input (I) and output (O) for a given chemical constituent tells whether that constituent is being accumulated within the ecosystem, $I > O$; is being lost from the system, $I < O$ or is quantitatively passing through the system, $I = O$. By this definition weathering products do not represent an input to the ecosystem; instead, weathering is considered as a process internal to the ecosystem and a part of the intrasystem cycle (Figure 1; Bormann and Likens, 1967; Likens and Bormann, 1972).

Annual Variation

We now have, depending on the chemical constituent, many years of annual budgets for precipitation input and stream-water output. We are therefore able to estimate with some accuracy mean annual budgets for most of the major ions (Tables 11 and 12). Over the long term there has been considerable variation; in general, however, a net loss of calcium, magnesium, potassium, sodium, sulfate, aluminum, and dissolved silica and a net gain of ammonium, hydrogen ion, and phosphate have occurred from these aggrading watershed ecosystems each year (Tables 11 and 12). The long-term averages suggest that there has been a small net gain in nitrate and chloride each year but there is no statistical difference ($p > 0.05$) between the annual inputs in bulk precipitation and stream-water outputs (Table 12).

There are two separate considerations to evaluating annual input–output budgets: (1) The *direction* of the net change, whether input is greater or less than output; and (2) the *magnitude* of the net change, i.e., the difference in amount between input and output. These parameters

TABLE 11. Annual Budgets of Bulk Precipitation Inputs and Stream-Water Outputs of Dissolved Substances for Undisturbed Watersheds of the Hubbard Brook Experimental Forest.

Substance (kg/ha)	1963–1964	1964–1965	1965–1966	1966–1967	1967–1968	1968–1969	1969–1970	1970–1971	1971–1972	1972–1973	1973–1974	Total (1963–1974), kg/ha	Annual mean, kg/ha
Calcium													
Input	3.0	2.8	2.7	2.7	2.8	1.6	2.3	1.5	1.2	1.2	2.0	23.8	2.2
Output	12.8	6.3	11.5	12.3	14.2	13.8	16.7	13.9	12.4	15.6	21.7	151.2	13.7
Net	−9.8	−3.5	−8.8	−9.6	−11.4	−12.2	−14.4	−12.4	−11.2	−14.4	−19.7	−127.4	−11.5
Magnesium													
Input	0.7	1.1	0.7	0.5	0.7	0.3	0.5	0.5	0.4	0.5	0.4	6.3	0.6
Output	2.5	1.8	2.9	3.1	3.7	3.3	3.5	3.1	2.8	3.3	4.6	34.6	3.1
Net	−1.8	−0.7	−2.2	−2.6	−3.0	−3.0	−3.0	−2.6	−2.4	−2.8	−4.2	−28.3	−2.5
Potassium													
Input	2.5	1.8	0.6	0.6	0.7	0.6	0.8	0.6	0.3	0.5	0.8	9.8	0.9
Output	1.8	1.1	1.4	1.7	2.2	2.2	2.1	1.7	1.6	2.1	2.8	20.7	1.9
Net	0.7	0.7	−0.8	−1.1	−1.5	−1.6	−1.3	−1.1	−1.3	−1.6	−2.0	−10.9	−1.0
Sodium													
Input	1.0	2.1	2.0	1.3	1.7	1.1	1.9	1.4	1.5	2.0	1.4	17.4	1.6
Output	5.9	4.5	6.9	7.3	9.1	7.6	8.0	6.5	5.8	8.1	9.8	79.5	7.2
Net	−4.9	−2.4	−4.9	−6.0	−7.4	−6.5	−6.1	−5.1	−4.3	−6.1	−8.4	−62.1	−5.6
Aluminum													
Input	[a]	[a]	[a]	[a]	[a]	[a]	[a]	[a]	[a]	[a]	[a]	[a]	[a]
Output	1.6[b]	1.2	1.7	1.9	2.1	2.2	2.2	1.8[b]	1.7[b]	2.3[b]	3.2[b]	21.9	2.0
Net	−1.6	−1.2	−1.7	−1.9	−2.1	−2.2	−2.2	−1.8	−1.7	−2.3	−3.2	−21.9	−2.0
Ammonium													
Input	2.6[b]	2.1	2.6	2.4	3.2	3.1	2.7	3.9	2.8	2.5	3.7	31.6	2.9
Output	0.27[b]	0.27	0.92	0.45	0.24	0.16	0.51	0.23	0.05	0.18	0.42	3.7	0.34
Net	2.3	1.8	1.7	2.0	3.0	2.9	2.2	3.7	2.8	2.3	3.3	27.9	2.6
Hydrogen													
Input	0.85[b]	0.76	0.85	1.05	0.96	0.85	0.93	1.18	0.97	1.08	1.14	10.62	0.96
Output	0.08[b]	0.06[b]	0.05	0.07	0.06	0.09	0.09	0.14	0.13	0.16	0.20	1.13	0.10
Net	0.77	0.70	0.80	0.98	0.90	0.76	0.84	1.04	0.84	0.92	0.94	9.49	0.86

Sulfate													
Input	33.7[b]	30.0	41.6	42.0	46.7	31.2	29.3	34.6	33.0	43.4	52.8	418.3	38.0
Output	42.7[b]	30.8	47.8	52.5	58.5	53.3	48.1	51.1	46.8	64.0	84.7	580.3	52.8
Net	−9.0	−0.8	−6.2	−10.5	−11.8	−22.1	−18.8	−16.5	−13.8	−20.6	−31.9	−162.0	−14.8
Nitrate													
Input	12.8[c]	6.7	17.4	19.9	22.3	15.3	14.9	21.6	21.4	26.3	30.9	209.5	19.0
Output	6.7[c]	5.6	6.5	6.6	12.7	12.2	29.6	24.9	18.7	19.2	34.8	177.5	16.1
Net	6.1	1.1	10.9	13.3	9.6	3.1	−14.7	−3.3	2.7	7.1	−3.9	32.0	2.9
Chloride													
Input	5.5[b]	4.5[b]	2.6[d]	6.2[b]	5.0	6.4	5.3	11.1	5.5	11.4	4.4	67.9	6.2
Output	3.7[b]	2.8[b]	4.3	4.8	5.3	5.2	4.2	4.6	3.6	6.0	5.9	50.4	4.6
Net	1.8	1.7	−1.7	1.4	−0.3	1.2	1.1	6.5	1.9	5.4	−1.5	17.5	1.6
Phosphate													
Input	0.09[b]	0.08[b]	0.10[b]	0.11[b]	0.11[b]	0.10[b]	0.10[b]	0.10[b]	0.10[b]	0.14	0.13	1.16	0.11
Output	0.02[b]	0.01[b]	0.02[b]	0.02[b]	0.02[b]	0.028	0.02[b]	0.02[b]	0.02[b]	0.018	0.029	0.225	0.020
Net	0.07	0.07	0.08	0.09	0.09	0.07	0.08	0.08	0.08	0.12	0.10	0.94	0.09
Bicarbonate[e]													
Input	[a]	[a]	[a]	[a]	[a]	[a]	[a]	[a]	[a]	[a]	[a]	[a]	[a]
Output	6.2[b]	4.6[b]	6.2	9.4	9.6	7.0	6.0	7.1[b]	6.6[b]	9.0[b]	12.5[b]	84.2	7.7
Net	−6.2	−4.6	−6.2	−9.4	−9.6	−7.0	−6.0	−7.1	−6.6	−9.0	−12.5	−84.2	−7.7
Dissolved silica (SiO_2)													
Input	[a]	[a]	[a]	[a]	[a]	[a]	[a]	[a]	[a]	[a]	[a]	[a]	[a]
Output	30.5[b]	20.8	36.1	41.6	42.1	35.0	41.4	33.8	30.6	45.7	56.7	414.3	37.7
Net	−30.5	−20.8	−36.1	−41.6	−42.1	−35.0	−41.4	−33.8	−30.6	−45.7	−56.7	−414.3	−37.7

[a]Not measured, but trace quantities.

[b]Calculated value, based on weighted average concentration during years when chemical measurements were made and on amount of precipitation or streamflow during the specific year.

[c]Calculated from weighted average concentration for 1964–1966 times precipitation or streamflow for 1963–1964.

[d]Based on annual concentration of 0.50 mg/l (Juang and Johnson, 1967).

[e]Watershed 4 only.

TABLE 12. Mean Annual (Arithmetic Average of Annual Values) Bulk Precipitation Input, Gross Stream-Water Output and Net Loss or Gain, and Standard Deviation of the Mean for Dissolved Substances, during 1963–1974 for the Hubbard Brook Experimental Forest.

	Input					Gross output					Net loss or gain[b]	
Substance	kg/ha	% of total	Eq × 10^3/ha	% of total	n^a	kg/ha	% of total	Eq × 10^3/ha	% of total	n^a	kg/ha	Eq × 10^3/ha
Ca^{2+}	2.17 ± 0.21	3	0.108 ± 0.010	8	11	13.7 ± 1.1	9	0.684 ± 0.055	42	11	−11.6 ± 1.2	−0.579 ± 0.060
Mg^{2+}	0.58 ± 0.07	1	0.048 ± 0.006	3	11	3.13 ± 0.21	2	0.257 ± 0.017	16	11	−2.55 ± 0.26	−0.210 ± 0.021
K^+	0.89 ± 0.19	1	0.023 ± 0.005	2	11	1.89 ± 0.14	1	0.048 ± 0.004	3	11	−0.99 ± 0.27	−0.025 ± 0.007
Na^+	1.59 ± 0.12	2	0.069 ± 0.005	5	11	7.23 ± 0.47	5	0.315 ± 0.020	19	11	−5.64 ± 0.50	−0.245 ± 0.022
Al^{3+}	—	—	—	—	—	1.87 ± 0.16	1	0.208 ± 0.018	13	6	−1.87 ± 0.16	−0.208 ± 0.018
NH_4^+	2.90 ± 0.18	4	0.161 ± 0.010	12	10	0.34 ± 0.08	<1	0.019 ± 0.004	1	10	+2.55 ± 0.21	+0.141 ± 0.012
H^+	0.977 ± 0.043	1	0.969 ± 0.043	70	10	0.114 ± 0.015	<1	0.103 ± 0.015	6	10	+0.872 ± 0.03	+0.865 ± 0.033
SO_4^{2-}	38.4 ± 2.5	52	0.799 ± 0.052	61	10	53.8 ± 4.4	36	1.12 ± 0.092	67	10	−15.3 ± 2.8	−0.319 ± 0.058
NO_3^-	19.7 ± 2.1	27	0.317 ± 0.034	24	10	17.1 ± 3.2	11	0.275 ± 0.052	17	10	+2.6 ± 2.7	+0.042 ± 0.044
Cl^-	7.01 ± 1.1	9	0.198 ± 0.032	15	7	4.89 ± 0.33	3	0.141 ± 0.009	8	7	+2.02 ± 1.10	+0.057 ± 0.031
PO_4^{3-}	0.131	<1	0.004	<1	2	0.023	<1	<0.001	<1	2	+0.108	+0.003
Dissolved silica	—	—	—	—	—	38.4 ± 3.0	26	—	—	10	−38.4 ± 3.0	—
HCO_3^-	—	—	—	—	—	7.64 ± 0.77	5	0.125 ± 0.013	8	5	−7.64 ± 0.77	−0.125 ± 0.013
Total	74.3	100	1.378 (cations) 1.318 (anions)			150.1	100	1.634 (cations) 1.661 (anions)			−75.8	−0.261 (cations) −0.342 (anions)

[a]number of years of measurement.

[b]There are small rounding errors associated with some of these means and their calculation from gross output minus input.

may vary from year to year for the individual chemical elements. For some ions, data from a single annual cycle ($n = 1$) are sufficient to establish the *direction* of net change; for other ions data from several years ($n > 1$) are necessary. Similarly, a variable number of years are required to characterize the *magnitude* of net change with meaningful confidence limits. Based on 11 yr of study at Hubbard Brook it is possible to make some generalizations about the annual budgets for the various ions. Generally, the direction of net change for budgets of precipitation input minus stream-water output for Ca^{2+}, Mg^{2+}, Na^+, Al^{3+}, NH_4^+, H^+, SO_4^{2-}, HCO_3^-, and dissolved silica was predictable from only a few years (1–3) of data (Table 11). For these ions the direction of the budget is not in doubt, but the magnitude may be. For K^+, NO_3^-, and CL^- one or a few years were likely to give unreliable results for both net change and magnitude. With the coefficient of variability (standard deviation of the mean divided by the mean net change; Table 12) used as a guide, a few annual measurements would yield reliable estimates of magnitude of loss of gain for Ca^{2+}, Mg^{2+}, Na^+, Al^{3+}, NH_4^+, H^+, SiO_2, and HCO_3^-. Nitrate, Cl^-, and K^+ also have a large range in annual magnitude of net change and they are much more unpredictable relative to direction of net change than the other ions (Table 11 and 12). Annual precipitation inputs for potassium and chloride have been particularly variable (Table 12). The standard deviation of the mean for long-term potassium inputs is 22%, and for chloride it is 17% of the mean. These highly variable inputs are therefore the major reason for the uncertainty in the annual budgets.

These data emphasize that for some elements data from one or a few annual cycles may give grossly aberrant estimates of long-term mean values of both direction and magnitude of change.

Sources of Input Other Than Bulk Precipitation

Nitrogen

During the 11-yr period 1963–1974, 209.5 kg/ha of nitrate (47.3 kg/ha NO_3-N) were added in bulk precipitation and 177.5 kg of NO_3^- per hectare (40.1 kg/ha NO_3-N) were lost in stream water from these forested, watershed ecosystems at Hubbard Brook. Likewise, 31.6 kg of ammonium per hectare (24.6 kg/hg NH_4-N) have been added in precipitation and 3.70 kg of NH_4^+ per hectare (2.9 kg/ha NH_4-N) have been lost in stream water during 1963–1974. This gives a total dissolved inorganic nitrogen input of 71.9 kg/ha and a nitrogen output of 43.0 kg/ha during this 11-yr period. These long-term data show that more inorganic nitrogen has been added to these forest ecosystems in precipitation than has been lost in stream water. This amount although substantial, cannot account for nitrogen accumulating within the ecosystem in living and dead biomass (Bormann et al., 1977; Covington, 1975, 1976; Whittaker et al., 1974).

Nitrogen is incorporated in this biomass and thereby is stored at a rate of about 16.7 kg/ha·yr in these forest ecosystems. Because nitrogen is essentially absent from the bedrock and till at Hubbard Brook, we assume that the release of nitrogenous substances by weathering and the formation of secondary nitrogenous minerals are negligible. Therefore, the net gaseous and aerosol exchange of nitrogen across the ecosystem's boundaries may be estimated (Figure 1) from net biomass accumulation minus net flux (precipitation input minus stream-water output). The long-term net gain in inorganinc nitrogen is 2.5 kg/ha·yr (precipitation input minus loss of dissolved and particulate nitrogen compounds in streamflow; Tables 10–12). The net accumulation in living and dead biomass is 16.7 kg/ha·yr. Therefore, ecosystem analysis using the small watershed technique shows that on the average about 14.2 kg of nitrogen per hectare·year must enter the ecosystem in a gaseous form or from impacted nitrogenous aerosols. Volatile exchanges of gases with the atmosphere have not been fully evaluated for the ecosystems at Hubbard Brook. Preliminary field data indicate that microbial fixation of nitrogen occurs within these forested ecosystems (Roskoski, 1975) and that denitrification, particularly chemical denitrification, may occur. It is likely that much of the additional accumulation of nitrogen within the ecosystem is largely a result of fixation of gaseous N_2 by microorganisms. If denitrification produced significant losses of gaseous nitrogen, then the actual amount of N_2 fixed would be even greater. Consequently, the actual input component for N (Tables 10 and 11) should be increased by some 14.2 kg/ha·yr, thereby increasing the average net gain value for N by a similar amount (Table 13).

Extensive work now has been done on all aspects of the nitrogen budget, and the above budget terms have changed as a result of these efforts. A monograph on the biogeochemistry of nitrogen at Hubbard Brook is in preparation.

Chlorine

With the exception of the relatively very large input of chloride in precipitation during 1970–1971 and 1972–1973, the long-term annual budget for chloride is nearly in balance. As mentioned above, long-term annual inputs are not statistically different than long-term outputs (Table 12). The high chloride input in 1970–1971 was largely a result of three or four exceptionally high concentrations in autumnal rainfall samples. These samples showed no evidence of contamination and repeated chemical analysis verified the results. Generally, the chloride budget at Hubbard Brook might be expected to be in balance (Juang and Johnson, 1967). We have no quantitative data on the biologic accumulation of chloride in these ecosystems but it is presumed to be small. However, substantial releases of chloride from some internal reservoir after disturbance by deforestation (Likens et al., 1970) suggest that chloride may be accumulating in the forest ecosystem. Chlorine, like nitrogen, may have a gaseous phase at normal biologic temperatures; however, chlorine gas is very reactive and is not common in the atmosphere. Significant amounts of chloride in throughfall and stemflow (Eaton et al., 1973; Table 7) suggested that small amounts of impacted aerosols containing chloride were washed off vegetation surfaces or that chloride was leached from vegetation by incident precipitation. Unfortunately, our data do not provide any definitive answers in this regard.

Sulfur

Sulfur also may exist as a gas or as an aerosol in the atmosphere. Storage of sulfur in the annual biomass increment occurs at a rate of about 2.0 kg/ha·yr. A small amount is released from chemical weathering (0.8 kg/ha·yr) within the ecosystems at Hubbard Brook, but the algebraic sum of the long-term precipitation input, stream-water output, weathering, and biomass accumulation shows that some 6.1 kg/ha·yr of sulfur must be accumulated each year from atmospheric gaseous and aerosol sources. This amount should be added to the bulk precipitation input value (Tables 11 and 12) for these ecosystems (Table 13). The origin,

Table 13. Summary of Annual Input–Output Budgets for Forested Ecosystems at Hubbard Brook.[a]

Element	Input, kg/ ha·yr	Output, kg/ha·yr	Net gain (+) or loss (−), kg/ha·yr
Si	—	23.8	−23.8
Ca	2.2	13.9	−11.7
Na	1.6	7.5	−5.9
Al	—	3.4	−3.4
Mg	0.6	3.3	−2.7
K	0.9	2.4	−1.5
Organic C	1484[b]	12.3	+1472
N	20.7	4.0	+16.7
Cl	6.2	4.6	+1.6
S	18.8	17.6	+1.2
H	0.96[c]	0.10[c]	+0.86
P	0.036	0.019	+0.017

[a]Input values for organic C, N, and S include bulk precipitation and estimates of dry deposition; all others are based on bulk precipitation only. Output values include both dissolved substances (Table 10) and particulate matter (Table 9) in stream water.

[b]Includes ecosystem biomass accretion (*net* ecosystem gaseous uptake of CO_2) based on the period 1961–1965 (Whittaker et al., 1974); therefore output does not include respiration losses of CO_2.

[c]Dissolved form only.

quantification, and ecological effects of gaseous or aerosol sulfur require detailed studies to elucidate the role of sulfur in forested ecosystems. Heretofore, these fluxes have been ignored or have not been quantitatively evaluated in biogeochemical studies.

Input–Ouput Budgets: Summary

Overall, during the period 1963–1974 there was an annual net loss of total dissolved inorganic substances from the experimental watersheds amounting to 74.7 kg/ha·yr (Table 11). The net loss averaged 261 Eq/ha·yr for cations and 342 Eq/ha·yr for anions (Table 12). The average net output of dissolved inorganic substances minus dissolved silica (1963–1974) was 37.1 kg/ha·yr. However, sizeable yearly variations have occurred during the study (Figures 15 and 21; Tables 11 and 12). For example, the smallest net loss of dissolved inorganic substances (27.8 kg/ha, or 7.0 kg/ha for total material minus dissolved silica) occurred during 1964–1965, the driest year of our study. The largest net losses of dissolved inorganic substances occurred during the very wet year, 1973–1974 (139.7 kg/ha and 986 cationic equivalents per hectare). Budgets adjusted to all sources of input are presented in Table 13. The explanation for the overall net losses lies in the biogeochemical reactions that occur within the ecosystem, noteably chemical weathering and related phenomena.

Seasonal Variations

Climate and biologic activity at Hubbard Brook are distinctly seasonal and much of the biogeochemistry reflects these climatic patterns. Normally the snowpack persists from about mid-November until April (Table 2). The trees start to produce leaves during late May, and the major leaf fall occurs in early October. Some organisms of the ecosystem have adapted to utilize efficiently the environmental conditions of the transition periods between seasons (cf. Muller and Bormann, 1976).

We have divided the year into four seasons to investigate the short-term variations in chemistry of precipitation and stream water. Summer conditions are assumed from 1 June to 30 September, autumn from 1 October to 30 November, winter from 1 December to 28 February, and spring from 1 March to 31 May. Although these periods are arbitrary, they correspond well with actual conditions and biologic activity at Hubbard Brook.

The weighted concentrations of various ions in precipitation are highly variable on a storm or weekly basis but some seasonal trends have emerged. The total ionic concentration of precipitation during the winter at Hubbard Brook is almost one-half that of other seasons of the year (Table 14). This is reflected in the concentration of individual ions, except

Table 14. Weighted Average Concentrations in Bulk Precipitation During Different Seasonal Periods for the Hubbard Brook Experimental Forest.

Substance	No. of years	June–Sept. mg/l	June–Sept. μEq/l	Oct.–Nov. mg/l	Oct.–Nov. μEq/l	Dec.–Feb. mg/l	Dec.–Feb. μEq/l	March–May mg/l	March–May μEq/l
Ca^{2+}	11	0.14	6.99	0.18	8.98	0.11	5.49	0.24	12.0
Mg^{2+}	11	0.04	3.29	0.07	5.76	0.03	2.47	0.05	4.11
K^{+}	11	0.07	1.79	0.11	2.81	0.04	1.02	0.07	1.79
Na^{+}	11	0.09	3.91	0.19	8.26	0.11	4.78	0.12	5.22
NH_4^{+}	10	0.24	13.3	0.19	10.5	0.15	8.32	0.27	15.0
H^{+}	10	0.090	89.3	0.070	69.4	0.045	44.6	0.079	78.4
SO_4^{2-}	10	3.72	77.5	2.34	48.7	1.67	34.8	3.23	67.2
NO_3^{-}	10	1.32	21.3	1.47	23.7	1.39	22.4	1.77	28.6
Cl^{-}	7	0.50	14.1	1.04	29.3	0.30	8.46	0.36	10.2
PO_4^{3-}	2	0.008	0.25	0.008	0.25	0.004	0.13	0.010	0.32
Total		6.22		5.67		3.85		6.20	
Σ^{+}			118.6		105.7		66.7		116.5
Σ^{-}			112.9		101.7		65.7		106.0

for sodium and nitrate. Concentrations of Na^{+} and NO_3^{-} during the winter are about equal to or greater than their concentrations during other seasons. A peak in concentration for chloride, sodium, potassium, and magnesium occurs in the autumn (October; Table 15). This annual sinusoidal pattern for many of the elements may result in part from the origin of air masses and in part from differences in aerosol scavenging efficiency between raindrops and snowflakes. Herman and Gorham (1957) have suggested from their studies in Nova Scotia that snowflakes may be less efficient than raindrops in removing materials from the atmosphere. Nitrate concentrations in precipitation are relatively similar during all seasons (Table 14). Other obvious but unexplained variations include: (1) Calcium equivalent concentrations are much greater than (almost double) sodium values during the spring and summer but about equal to sodium during the autumn and winter; and (2) the equivalent concentration of chloride exceeds that of nitrate during autumn but nitrate concentrations greatly exceed chloride values during the other seasons. The relatively high concentrations of sodium and chloride in autumn rains may reflect a larger influence from marine aerosols at this time in the eastern United States. However, chloride:sodium ratios during summer and autumn are identical (5.6 on a weight basis) and are far in excess of those expected (1.8) from seawater (Junge, 1963). The high Cl:Na ratios during summer and autumn at Hubbard Brook may indicate some seasonal transport of the chloride to the area from distant anthropogenic sources (Gorham, 1958; Junge, 1963).

Concentrations of various ions in stream water are much less variable seasonally than in precipitation (Table 16). Concentrations in stream

TABLE 15. Monthly Bulk Precipitation and Stream-Water Values for the Hubbard Brook Experimental Forest During 1963–1974.[a]

		Water[b]	Ca^{2+}[b]	Mg^{2+}[b]	K^{+}[b]	Na^{+}[b]	H^{+}[c]	NH_4^{+}[c]	NO_3^{-}[c]	SO_4^{2-}[c]	Cl^{-}[d]	PO_4^{3-}[e]
Bulk precipitation												
Summer	June	12.07	0.15	0.05	0.10	0.13	0.0787	0.23	1.44	3.8	0.83	0.007
	July	11.07	0.17	0.05	0.07	0.07	0.1054	0.27	1.46	4.4	0.25	0.010
	August	13.38	0.13	0.03	0.04	0.06	0.0751	0.23	1.24	3.2	0.39	0.012
	September	9.56	0.12	0.04	0.06	0.08	0.1041	0.23	1.53	3.5	0.44	0.007
Autumn	October	8.63	0.25	0.08	0.13	0.23	0.0786	0.23	1.69	2.6	1.00	0.013
	November	13.58	0.14	0.07	0.09	0.17	0.0644	0.16	1.54	2.2	1.07	0.005
Winter	December	14.93	0.09	0.03	0.05	0.09	0.0432	0.11	1.37	1.6	0.30	0.005
	January	7.96	0.10	0.02	0.04	0.13	0.0528	0.23	1.89	2.1	0.29	0.005
	February	9.29	0.15	0.04	0.03	0.14	0.0423	0.15	1.37	1.4	0.32	0.001
Spring	March	10.25	0.20	0.04	0.04	0.12	0.0630	0.21	1.94	2.5	0.33	0.003
	April	9.86	0.27	0.06	0.06	0.11	0.0779	0.31	1.88	3.1	0.31	0.013
	May	11.67	0.25	0.05	0.10	0.11	0.0935	0.30	1.88	3.9	0.42	0.013
Stream water												
Summer	June	4.91	1.46	0.32	0.16	0.84	0.0127[f]	0.02	0.54	6.6	0.38	0.002
	July	2.28	1.54	0.33	0.14	0.87	0.0135[f]	0.04	0.45	6.8	0.37	0.002
	August	1.66	1.61	0.37	0.14	1.01	0.011[f]	0.09	0.39	7.1	0.50	0.004
	September	1.58	1.57	0.40	0.11	1.17	0.0082[f]	0.05	0.27	7.4	0.57	0.003
Autumn	October	3.83	1.58	0.40	0.22	1.01	0.0086[f]	0.06	0.31	7.0	0.56	0.002
	November	7.92	1.74	0.41	0.23	0.97	0.0098[f]	0.09	1.20	6.8	0.70	0.001
Winter	December	8.11	1.71	0.39	0.20	0.89	0.0131[f]	0.04	2.46	6.3	0.49	0.002
	January	3.86	1.69	0.39	0.20	0.96	0.0118[f]	0.03	2.17	6.4	0.54	0.002
	February	3.66	1.74	0.41	0.21	0.98	0.0107[f]	0.05	2.72	6.2	0.51	0.001
Spring	March	9.32	1.71	0.40	0.24	0.92	0.0115[f]	0.04	2.80	6.3	0.54	0.001
	April	23.85	1.69	0.38	0.26	0.77	0.0135[f]	0.03	2.95	6.1	0.51	0.002
	May	12.11	1.56	0.34	0.25	0.79	0.0143[f]	0.02	1.67	6.2	0.46	0.002

[a]Mean weighted concentrations of dissolved substances given in mg/l; water given as cm/unit area.
[b]1963–1974.
[c]1964–1974.
[d]1967–1974.
[e]1972–1974.
[f]1965–1974.

TABLE 16. Weighted Average Concentrations in Stream Water during Different Seasonal Periods for Forest Watersheds of the Hubbard Brook Experimental Forest.

Substance	No. of years	June–Sept. mg/l	June–Sept. μEq/l	Oct.–Nov. mg/l	Oct.–Nov. μEq/l	Dec.–Feb. mg/l	Dec.–Feb. μEq/l	March–May mg/l	March–May μEq/l
Ca^{2+}	11	1.49	74.4	1.69	84.3	1.71	85.3	1.66	82.8
Mg^{2+}	11	0.34	28.0	0.41	33.7	0.40	32.9	0.37	30.4
K^+	11	0.14	3.58	0.23	5.88	0.20	5.11	0.25	6.39
Na^+	11	0.91	39.6	0.98	42.6	0.92	40.0	0.81	35.2
NH_4^+	10	0.04	2.22	0.08	4.44	0.04	2.22	0.03	1.66
H^+	9	0.012	11.9	0.009	8.93	0.012	11.9	0.013	12.9
SO_4^{2-}	10	6.7	139	6.8	142	6.3	131	6.2	129
NO_3^-	10	0.44	7.10	0.89	14.4	2.46	39.7	2.57	41.5
Cl^-	10	0.49	13.8	0.62	17.5	0.57	16.1	0.54	15.2
PO_4^{3-}	2	0.003	0.09	0.001	0.03	0.002	0.06	0.002	0.06
Total		10.6		11.7		12.6		12.4	
Σ^+			158.7		179.8		177.4		169.3
Σ^-			160.0		173.9		186.9		185.8

water are slightly lower in the summer, in part because of the intense biologic utilization of potassium and nitrate during the growing season.

In general, nitrate concentrations in stream water begin to rise in the autumn when there is a reduction in biologic activity, and attain a maximum concentration in the winter or early spring. United States Geological Survey Water Resources data show a similar pattern for rivers and streams throughout New England and New York State. This pattern also has been observed in the surface waters of many freshwater lakes where nitrate concentrations reach maximum levels in winter and early spring. This latter increase in nitrate concentrations is attributed to increased nitrification during the winter (Hutchinson, 1957, p. 871).

At Hubbard Brook we have observed that nitrate concentrations increase markedly with increased streamflow (Johnson et al., 1969). However this may be a fortuitous relationship since the late winter-early spring period of maximum streamflow is potentially also a time for relatively high nitrification in the soil. Even though there may be no causal relationship between high nitrate concentration and increased streamflow during this time, the fact that this period of high nitrate concentrations also is characterized by large discharge remains of prime importance to nutrient budget calculations.

The seasonal pattern for ammonium concentrations in stream water is somewhat unique. The clear peak in ammonium concentration during the autumn may reflect increased decomposition of organic matter in the forest floor after leaf fall.

The pattern of average precipitation input and stream-water output during the seasonal periods (Tables 17 and 18) similar to the monthly flux (Figures 35–39), but the resolution is not as great. Input (on a weight basis) is greatest during the summer and least during the autumn (Table 17). The equivalent input is lowest during the winter, largely because of the sharp decline in chloride inputs. On a weight basis, sulfate and nitrate provide about 70–80% of the dissolved inorganic substances in precipitation. On an equivalent basis, hydrogen ion and sulfate dominate, as expected. In general the input of nitrate reflects more closely than any other ion the amount of water added as precipitation (Table 17). Inputs of magnesium, potassium, and to a lesser extent calcium and sodium are relatively greater than expected on the basis of amount of precipitation during the autumn (enriched precipitation relative to the annual average) and less during the winter. All ionic inputs during the winter season were less than expected on the basis of amount of precipitation; in general, however, inputs of individual dissolved substances were highly variable on a seasonal basis (Table 17).

More than 55% of the annual output of dissolved substances in stream water occurs during the spring. In contrast, only 11% of the output occurs during summer (Table 18). Sulfate clearly dominates the output of dissolved substances on a weight and equivalent basis. Calcium, magnesium, sulfate, chloride, and, to a lesser extent, hydrogen ion and sodium losses in stream water reflect the amount of water lost. Phosphate losses bear little relationship to the amount of streamflow; summer and winter losses are appreciably greater than expected and autumn and spring losses are less than expected from the amount of streamflow. Likewise, output of nitrate during summer and autumn is less than expected and during winter and spring is more than expected on the basis of water loss. Outputs of potassium and ammonium are variable, but the reduction of potassium during the summer produces an obvious drop in stream-water output assumed to be from biologic utilization within the forested ecosystem (Table 18).

Monthly Variations

A greater insight into the biogeochemistry of these forest ecosystems can be gained by looking at the nutrient budgets on a monthly scale. Three patterns of input–output emerge: those in which input exceeds output in every month, those in which output exceeds input in every month, and those in which there is a crossover from input dominance to output dominance on a monthly basis. These patterns are not related to whether the elements have a gaseous component or are entirely sedimentary. For example, Ca, P, and K are all sedimentary, yet they exhibit different patterns; S and N, however, both atmosphilic, exhibit two patterns.

TABLE 17. Average Precipitation Input during Different Seasonal Periods for Forest Watersheds of the Hubbard Brook Experimental Forest.

Substance	No. of years	June–September g/ha	June–September Eq/ha	June–September % of annual total	October–November g/ha	October–November Eq/ha	October–November % of annual total	December–February g/ha	December–February Eq/ha	December–February % of annual total	March–May g/ha	March–May Eq/ha	March–May % of annual total
Ca^{2+}	11	656	32.7	30.2	405	20.2	18.7	343	17.1	15.8	767	38.3	35.3
Mg^{2+}	11	182	15.0	31.3	153	12.6	26.4	89.6	7.37	15.4	156	12.8	26.9
K^{+}	11	310	7.93	34.7	238	6.09	26.6	126	3.22	14.1	220	5.63	24.6
Na^{+}	11	393	17.1	25.3	431	18.7	27.7	361	15.7	23.2	369	16.1	23.7
NH_4^{+}	10	1,140	63.2	39.3	413	22.9	14.2	491	27.2	16.9	859	47.6	29.6
H^{+}	10	425	422	43.5	155	154	15.9	149	148	15.3	248	246	25.4
SO_4^{2-}	10	17,600	366	45.9	5,170	108	13.5	5,500	115	14.3	10,100	210	26.3
NO_3^{-}	10	6,270	101	31.9	3,240	52.3	16.5	4,580	73.9	23.3	5,550	89.5	28.3
Cl^{-}	7	2,380	67.1	34.0	2,290	64.6	32.7	1,110	31.3	15.8	1,230	34.7	17.5
PO_4^{3-}	2	51.1	1.61	39.0	20.0	0.632	15.3	19.1	0.603	14.6	40.7	1.29	31.1
Total	—	29,356	$\Sigma + 558$ $\Sigma - 536$	39.5	12,515	$\Sigma + 234$ $\Sigma - 226$	16.8	12,769	$\Sigma + 219$ $\Sigma - 221$	17.2	19,540	$\Sigma + 366$ $\Sigma - 335$	26.3
Water (cm/ha)	11	46.0		34.8	22.2		16.8	32.2		24.4	31.8		24.1

TABLE 18. Average Stream-Water Output During Different Seasonal Periods for Forest Watersheds of the Hubbard Brook Experimental Forest.

		June–September			October–November			December–February			March–May		
Substance	No. of years	g/ha	Eq/ha	% of annual total	g/ha	Eq/ha	% of annual total	g/ha	Eq/ha	% of annual total	g/ha	Eq/ha	% of annual total
Ca^{2+}	11	1,580	78.8	11.5	1,980	98.8	14.4	2,680	134	19.5	7,510	375	54.6
Mg^{2+}	11	357	29.4	11.4	476	39.2	15.1	619	50.9	19.7	1,690	139	53.8
K^+	11	151	3.86	8.0	268	6.85	14.2	316	8.08	16.8	1,150	29.4	61.0
Na^+	11	965	42.0	13.3	1,160	50.5	16.0	1,450	63.1	20.0	3,660	159	40.6
NH_4^+	10	45.0	2.49	13.4	96.6	5.36	28.7	57.4	3.18	17.0	138	7.65	40.9
H^+	9	14.4	14.3	13.2	11.9	11.8	10.9	19.6	19.4	17.9	63.4	62.9	58.0
SO_4^{2-}	10	7,740	161	14.3	8,240	172	15.3	9,940	207	18.4	28,100	585	52.0
NO_3^-	10	513	8.27	3.0	1,070	17.3	6.2	3,880	62.6	22.6	11,700	189	68.2
Cl^-	10	569	16.1	12.2	746	21.0	15.9	903	25.5	19.3	2,460	69.4	52.6
PO_4^{3-}	2	6.72	0.212	29.1	1.72	0.054	7.4	5.80	0.183	25.1	8.89	0.281	38.4
Total	—	11,941	$\Sigma + 171$ $\Sigma - 186$	11.7	14,050	$\Sigma + 213$ $\Sigma - 210$	13.7	19,871	$\Sigma + 279$ $\Sigma - 295$	19.4	56,480	$\Sigma + 773$ $\Sigma - 843$	55.2
Water (cm/ha)	11	10.6		12.7	11.7		14.1	15.6		18.7	45.3		54.4

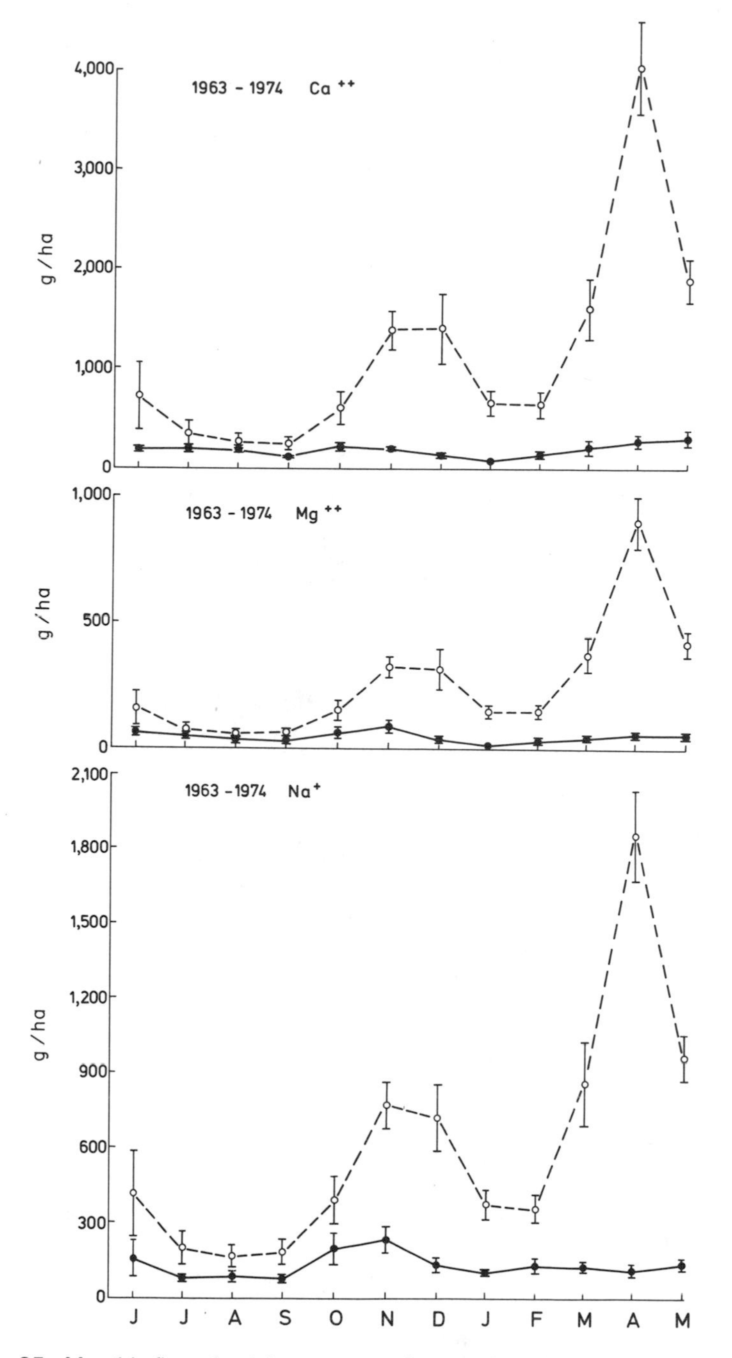

FIGURE 35. Monthly flux of calcium, magnesium, and sodium for forest watershed ecosystems of the Hubbard Brook Experimental Forest. Vertical bars indicate ± one standard deviation of the mean. Solid line is input and dashed line is output.

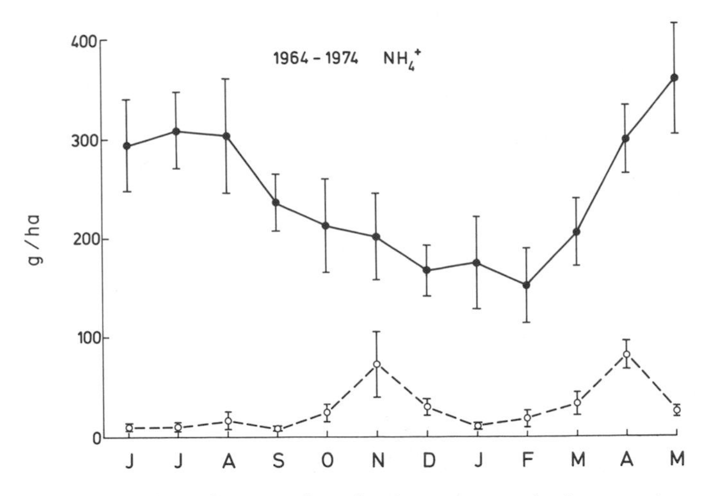

FIGURE 36. Monthly flux of ammonium for forest watershed ecosystems of the Hubbard Brook Experimental Forest. Vertical bars indicate ± one standard deviation of the mean. Solid line is input and dashed line is output.

Instead of this simple assumption, a complex interaction involving such factors as input, effect of biologic activity, and annual variations in weather are determining factors.

Output Consistently Greater Than Input

Magnesium exlemplifies a nutrient in which the output exceeds the input each month of the year on a long-term basis (Figure 35). The effects of increased streamflow are clearly shown by the peak chemical outputs in April and November-December. During a specific year, the monthly input and output may be quite variable, and monthly outputs may not exceed inputs. In fact, based on a 6-yr average (1963–1969) inputs exceeded the outputs during July. Calcium, sodium, and aluminum all have patterns similar to magnesium; i.e., outputs consistently exceed inputs on a monthly basis (Figure 25).

Input Consistently Greater Than Output

In sharp contrast to magnesium, ammonium is characterized by monthly inputs that consistently exceed outputs on a long-term basis (Figure 36). Monthly outputs of ammonium are relatively constant and low. The pattern of ammonium inputs may be highly variable for a specific year but it becomes much clearer and distinctly sinusoidal with long-term data. The monthly input during late autumn and winter is relatively low. This may reflect a reduced rate of ammonia input into the atmosphere from decay in the soil because of the lower temperatures, or the transfer of ammonia to the atmosphere may be reduced by the presence of a snowpack. Likewise, higher inputs during the spring and summer may reflect the general increase in biologic activity of the soil and the elevated generation of ammonia. However, because ammonia may be converted rapidly to nitrate, N_2O, N_2, etc., it is difficult to interpret these monthly fluxes. The overall monthly fluxes for hydrogen and phosphate are similar to ammonium (input exceeds output).

Crossover Patterns

Potassium inputs exceed outputs from mid-May until mid-October and during the remainder of the year monthly outputs are larger than inputs (Figure 37). This is of interest, for we have suggested earlier that the potassium budget (particularly potassium in stream water) is particularly sensitive to seasonal biologic control within the ecosystem (Tables 16 and 18). As mentioned above, the forest vegetation of these watersheds normally breaks dormancy about mid-May and the deciduous leaves are shed about mid-October. Apparently the biotic portion of the ecosystem actively extracts potassium from the soil and drainage waters during the growing season (cf. Johnson et al., 1969). The "crossover times" separating input or output domination were not clearly defined in the "long-term" monthly budgets by 1969 (1963–1969), which presumably was a residual effect of the drought years (1963–1965) on the cumulative average (see Table 11). The large monthly export of potassium in stream water during March through May of each year generally surpasses the average annual input to these forest ecosystems (Figure 37; Tables 17 and 18). The monthly pattern of potassium inputs and outputs for any specific year may be highly variable. For example, as mentioned above, inputs exceeded outputs during the drought years of 1963–1965 (Figure 38) but less dramatic, although complicated, patterns occurred in other years.

The monthly pattern for sulfate is generally similar to that of potassium (Figure 37) except that the long-term results show that the crossover from input to output dominance in the autumn occurs somewhat earlier (late September) than for potassium. The monthly input of sulfate in bulk precipitation has a rather symmetrical, sinusoidal pattern, with a peak in July and a low in January–February. During the summer and early autumn of 1968 the input was unusually small and during April the output was unusually large, both of which combined to produce the largest annual net output for any year except the very wet year of 1973–1974 (Table 11).

Monthly nitrate flux requires special consideration. The monthly pattern is generally consistent but distinctly different than the pattern for ammonium (Figures 36 and 39), although the annual patterns for these two

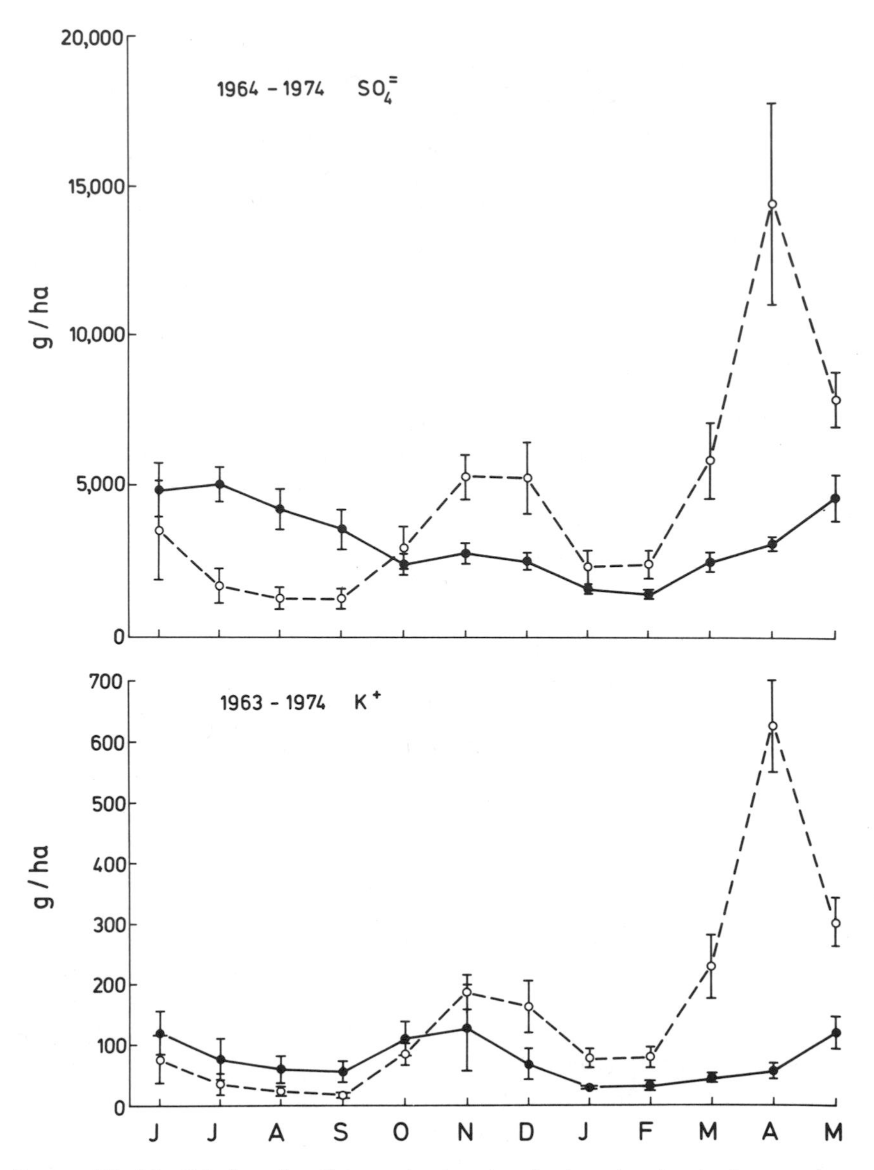

FIGURE 37. Monthly flux of sulfate and potassium for forest watershed ecosystems of the Hubbard Brook Experimental Forest. Vertical bars indicate ± one standard deviation of the mean. Solid line is input and dashed line is output.

nitrogenous ions are similar (i.e., inputs exceed outputs). The long-term pattern for nitrate after 5 yr (1964–1969) and after 8 yr (1964–1972) were similar and showed that inputs exceeded outputs during every month except April, even though after 8 yr the average output for April was

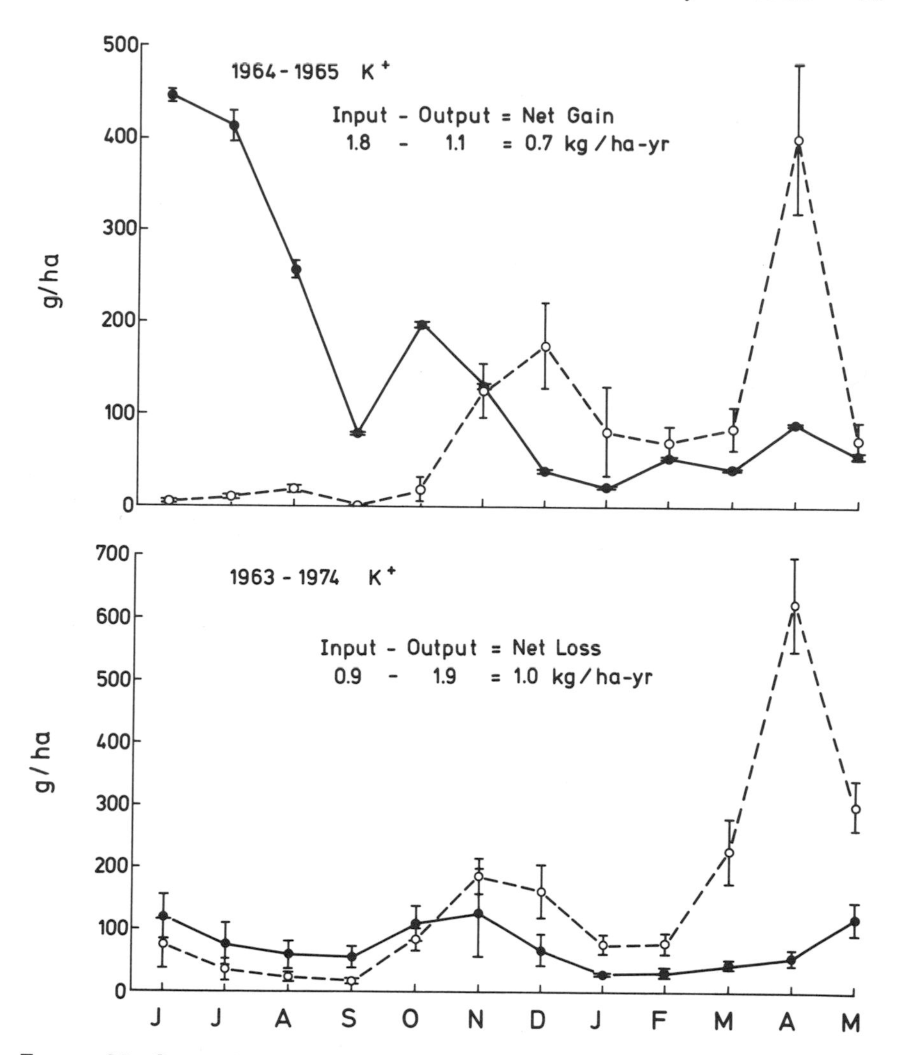

FIGURE 38. Comparison of the monthly flux of potassium during the drought years of 1963–1965 with the long-term (1963–1974) average. Vertical bars indicate ± one standard deviation of the mean. Solid line is input and dashed line is output.

55% higher than it was after only 5 yr. However, during 1969–1970, 1970–1971, and 1973–1974 a distinct change in the annual pattern occurred (Table 11). During these 3 yr the long-term annual relationship, where input exceeds output, was reversed. The input during August, September, October, November, and January 1969–1970 was less than the long-term average but it was the relatively large outputs in April and to a lesser extent in December and February that overwhelmed the

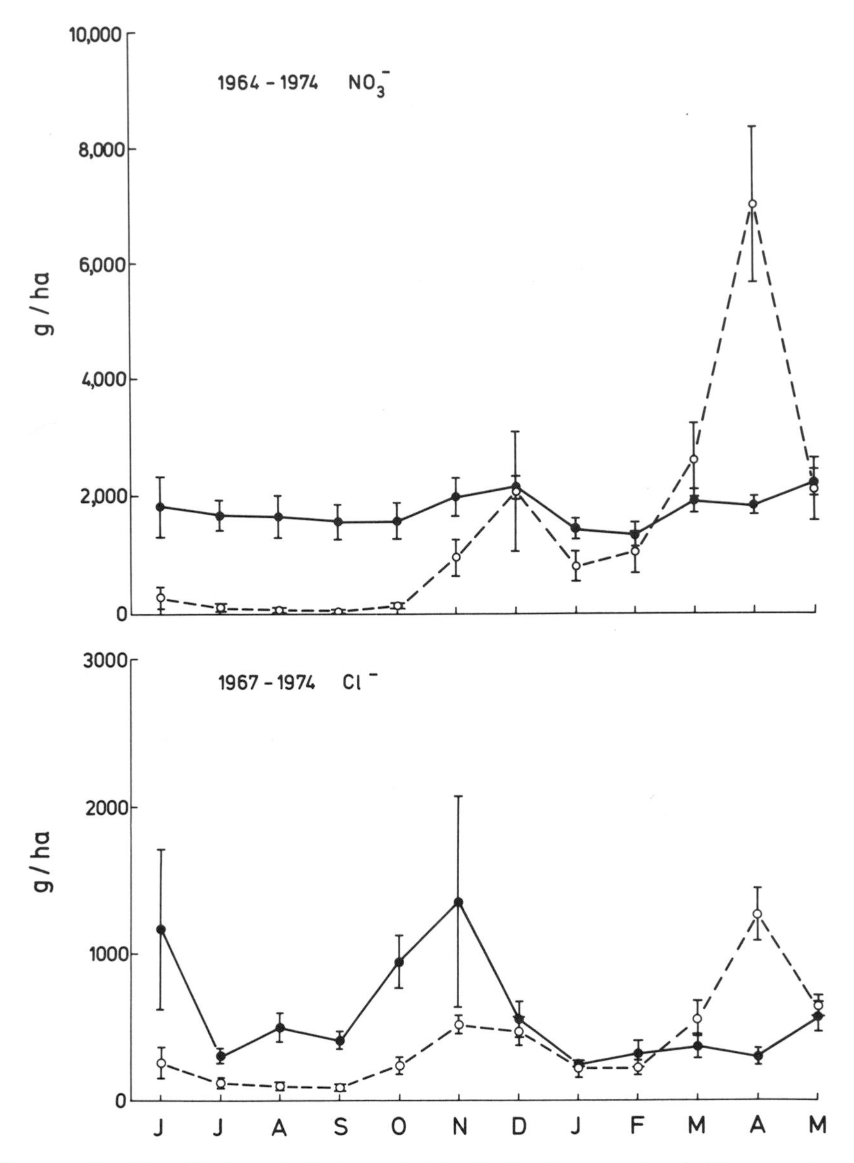

FIGURE 39. Monthly flux of nitrate and chloride for forest watershed ecosystems of the Hubbard Brook Experimental Forest. Vertical bars indicate ± one standard deviation of the mean. Solid line is input and dashed line is output.

longterm annual budget pattern (Figure 39). As mentioned previously this change in pattern during 1969–1970 and 1973–1974 may have been related to the unusual occurrence of soil frost.

The trend of increasing concentration (and input) of nitrate in precipitation during the period of study (Figures 9 and 17; Table 11) also has affected the long-term monthly flux. As a result of these perturbations in precipitation and stream-water flux, the resolution of the "average" long-term monthly pattern for nitrate is not good. Nevertheless, based on 10 yr of data, monthly inputs of nitrate exceed outputs during all months except December, March, April, and May. Inputs and outputs are essentially balanced during December and May (Figure 39).

The long-term monthly pattern for chloride is similar to that for nitrate (Figure 39).

Addendum

Because of long-term changes in the chemistry of precipitation and stream water, there have been associated changes in the input–output budgets for the watershed ecosystems at Hubbard Brook. As an example, data are presented in Table 19 for Watershed 6 during 1992–1993. This water-year was relatively typical in terms of amount of precipitation as compared to the long-term average. In general, the overall patterns for the input/output relations have remained remarkably the same with time, but there are some important differences in detail. The watershed-ecosys-

TABLE 19. Volume-Weighted Concentration (μeq/liter) and Flux (eq/ha·yr) of Major Ions and Dissolved Silica during 1992–1993 for Watershed 6 of the Hubbard Brook Experimental Forest.

	Bulk precipitation			Stream water		
	Concentration			Concentration		
Dissolved substance	(μeq/liter)	% total cations/anions	Input flux eq/ha-yr	(μeq/liter)	% total cations/anions	Output flux eq/ha-yr
H^+	52.5	69	720	12.0	9	107
NH_4^+	12.2	16	167	1.7	1	15.1
Ca^{2+}	3.2	4	43.9	39.5	29	351
Na^+	4.6	6	63.1	29.1	21	259
Mg^{2+}	1.8	2	24.7	18.5	13	165
K^+	1.1	1	15.1	4.2	3	37.4
Al^{3+}	0.9	1	12.3	32.2	23	286
SO_4^{2-}	40.5	56	556	97.0	87	863
NO_3^-	26.1	36	358	3.6	3	32.0
Cl^-	5.7	8	78.2	10.5	9	93.4
Dissolved Si (mg/liter)		0.05	—		3.69	—
pH		4.28	—		4.92	—
Water (mm)		1372			889.7	
$\Sigma+$	76.3				137	
$\Sigma+$	72.3				111	

tems continue to show significant net gains (precipitation input minus streamflow output) for inorganic nitrogen, hydrogen ion and sulfur (including dry deposition input). It is interesting to note that currently, stream-water concentrations of nitrate in Watershed 6 are the lowest on record (Figure 29), even though forest biomass accumulation is very low. Significant net losses of calcium, magnesium, sodium and potassium occur, but stream-water outputs of calcium, magnesium and sodium are appreciably smaller than during 1963–1974, and as a percentage of total cation loss, have become smaller (Tables 12 and 19). In contrast, dissolved aluminum is much more important in the stream-water flux, where now it is the second most dominant cation; in 1963–1974 it was 4th (Tables 12 and 19). During 1992–1993 there was a net loss of chloride (Table 19), but the input/output balance for this ion has varied from year (see Table 12).

The balance between the sum of cations and the sum of anions in Table 19 is not good, but does not include bicarbonate or dissolved organic anions and the valence used for dissolved aluminum was +3. Making such adjustments would reduce the discrepancy to a small error.

5 Weathering

Both the qualitative and the quantitative changes in water chemistry elicited during the passage of water through the ecosystem are related in part to the process of chemical weathering. The bulk ionic composition of water entering the Hubbard Brook ecosystem is essentially characterized by acid salts, such as H_2SO_4, HNO_3, and HCl. In contrast, water leaving the system is characterized mainly by neutral salts, such as $CaSO_4$, Na_2SO_4, $Mg(NO_3)_2$, and to a lesser extent by chlorides and bicarbonates. This qualitative change in chemistry exemplifies the general chemical weathering reactions. This reaction and some of the chemical pathways in an open system, such as the Hubbard Brook watershed ecosystem, are shown in Figure 40.

As H^+ is fixed within the system, basic cations (M^+) may be leached from the systems and primary silicate minerals (M^+X) are transformed into clays or hydroxides and left behind as a residuum (Figure 40). When mechanical erosion or other disturbance is sufficiently slow the mineral residuum interacts with organic substances to form soil.

The rate at which the chemical weathering reaction proceeds may be seen as the rate at which H^+ is supplied to the system (see Figure 40). An alternate viewpoint, however, for determining weathering rate is from the rate at which cations are withdrawn from the system, i.e., cationic denudation (Reynolds and Johnson, 1972).

Sources of Hydrogen Ion

The hydrogen ion, which drives the chemical weathering reaction, is supplied from two sources at Hubbard Brook, one external to the ecosystem and the other internal. The external source is acids supplied in bulk precipitation per se (meteorologic input); the internal source is various biologic and chemical processes occurring within the soil zone. The meteorologic input of H^+ at Hubbard Brook is mainly in the form of H_2SO_4 and HNO_3 (Tables 11 and 12). The average external net input of

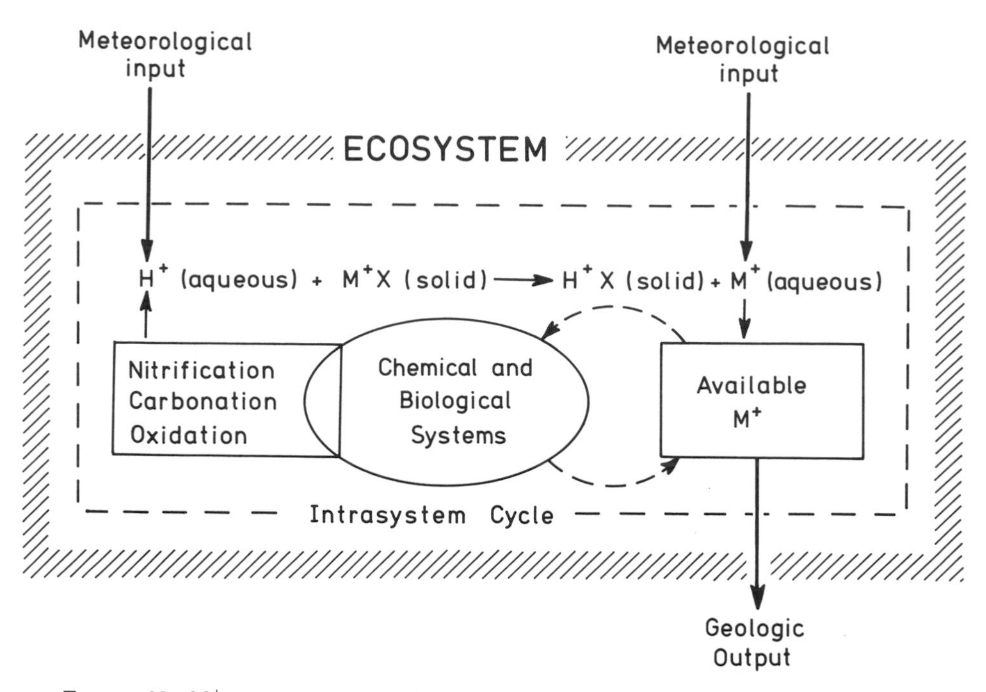

FIGURE 40. M^+ represents a basic cation and X is an ionic exchange substrate, usually a primary silicate mineral, but also includes such derivatives as soil minerals or soil humates. The above illustrates that as H^+ is added to the system from external and internal sources, basic cations (M^+) are leached from the system, and primary silicate minerals (M^+X) are transformed into secondary minerals (H^+X).

hydrogen ion equivalent observed at Hubbard Brook over the past decade is $0.865 \pm .033 \times 10^3$ Eq/ha·yr (Table 12). If this were the only source of H^+ ions at Hubbard Brook (and the ecosystem were in a steady state), we might expect this value for hydrogen fixation to be more or less balanced by the net rate at which ionic Ca, Mg, K, Na, and Al were leached from the system. In fact, there are more of these cations removed from the ecosystem each year ($1.267 \pm 0.071 \times 10^3$ Eq/ha) than there are external hydrogen ions to replace them (Table 12). The difference ($0.402 \pm .078 \times 10^3$ Eq/ha·yr) is statistically significant and implies the action of internally generated H^+. By this definition "internal" is taken to mean everything that is not sensibly added by meteorologic input.

Some of the internal sources of H^+ at Hubbard Brook are as follows.

Nitrogen

An interesting aspect of the chemical weathering reaction at Hubbard Brook is the role of the ammonium ion. It is well known to agriculturalists that additions of ammonium ion have an acidifying effect on soils (Brady, 1974; Reuss, 1976). When ammonium ion is fully oxidized to nitrate ion two hydrogen ions are produced. Ammonium is added to the ecosystem both by meteorologic input and by decomposition within the system. The actual acidifying effect of NH_4^+ in the Hubbard Brook soils is unknown; however, it is estimated that only about 10 to 20% of the soil ammonium is oxidized (Melillo, 1997). The rest is apparently taken up directly by plants, which may release additional H^+ (Reuss, 1976). Although nitrification represents a substantial source of H^+, it is still not sufficient by itself to make up the observed deficit between cationic output and external H^+ input.

Carbon

Reduced carbon in the form of CHO compounds is continually oxidized in the soil zone with the production of CO_2 at partial pressures usually exceeding those found in the atmosphere. Some CO_2 reacts with the soil water to form carbonic acid, which has a potential to generate H^+ ions. However, the soils at Hubbard Brook, at least in the upper horizons, are quite acid ($pH < 5$). Under these conditions H_2CO_3 dissociates only very slightly and hence carbonation weathering reactions (Carroll, 1970) are inhibited. An additional potential source of H^+ at Hubbard Brook is from the production of organic acids, such as citric, tartaric, tannic, and oxalic acids, by biologic activity within the soil zone. The role of such organic acids in chemical weathering is undoubtedly important in the Hubbard Brook ecosystem. However, the quantitative effects that these organic acids exert cannot be assessed with our present experimental data.

Sulfur

There are small amounts of sulfide minerals in the bedbrock at Hubbard Brook that are subject to oxidation in the soil zone with the concommittant production of sulfuric acid (Carroll, 1970). In addition microbially decomposed organic sulfur compounds, such as sulfur, sulfides, thiosulfates and polythionates, are subject to oxidation with the production of sulfuric acid or sulfates. Oxidation may occur by strict chemical reactions but most sulfur oxidation occurring in the soil profile is thought to be biochemical in nature (Brady, 1974).

Estimates of Weathering Rates

As mentioned above the long-term cationic denudation in the Hubbard Brook ecosystem is measured at $1.267 \pm .071 \times 10^3$ Eq/ha·yr. This situation is somewhat more complicated, however, because we know that both living and dead biomass is accumulating within the forest ecosystems at Hubbard Brook. In effect, this net accretion of biomass represents a long-term sink for some of the nutrients supplied from the weathering reaction. The total amount of cations sequestered by this means is 0.722 $\times$ 10^3 Eq/ha·yr.

This disposition of cations within the ecosystem allows some interesting conclusions. (a) Cations stored within the biomass must be included in calculations of contemporary weathering; (b) the rate of storage is a consequence of the ecosystem's current state of forest succession and so changes with time; and (c) existence of the forest and its state of development must be included in geologic estimates of weathering.

Based on biomass considerations, reported more fully in our second volume, we can estimate the role of biomass accumulation in weathering. As indicated previously the HBEF is not presently in a steady state with respect to biomass. It is recovering from a logging operation some 75 yr ago (in 1994) and currently is accumulating biomass slowly, which includes a cationic component. Until such time that steady-state conditions do pertain within the forest some of the nutrients provided by chemical weathering must continue to be taken up and stored by the biologic system (Johnson, 1971). Then it follows that the flux of cationic nutrients being diverted into biomass accretion (0.722 $\times$ 10^3 Eq/ha·yr) must be added to that actually removed from the system in the form of dissolved load (1.267 $\times$ 10^3 Eq/ha·yr) and particulate organic matter ($\sim$0.01 $\times$ 10^3 Eq/ha·yr). Therefore, our best estimate of cationic denudation (net loss from ecosystem plus long-term storage within the system) at Hubbard Brook becomes about 2.0 $\times$ 10^3 Eq/ha·yr during 1963–1974.

Because the cationic denudation values presented here include biomass accumulation, they are about twice those reported previously (Johnson et al., 1972). For comparison purposes, it is noteworthy that the annual cationic denudation we have determined for Hubbard Brook (2.0 $\times$ 10^3 Eq/ha) is comparable to that calculated for New England as a whole (2.2 $\times$ 10^3 Eq/ha) but substantially less than the average (3.8 $\times$ 10^3 Eq/ha) for the North American continent (Reynolds and Johnson, 1972).

Although it is interesting to note the essential concordance in cationic denudation rates between Hubbard Brook and values for the New England region as a whole, the reason for this concordance is not so clear. The Hubbard Brook data relate to an upland stream system that is sensibly unpolluted and that has a basin composed of igneous and metamorphic rocks. The cationic denudation here is strictly derived from chemical weathering activity. In contrast, the regional values of cationic denudation have been estimated from the large, lowland streams of New England, which flow through long reaches of urban areas, and sedimentary rocks. Ostensibly, these streams have abundant opportunities to pick up dissolved loads of road salt, pollutants, and congruently soluble natural salts, all of which do not reflect chemical weathering activity per se. The similarity then between the cationic denudation values for Hubbard Brook and the New England region may be fortuitous, relating as they do to entirely different sets of chemical provenance and pollutive conditions. The similarity also begs an intriguing question: Why is the cationic denudation of lowland basins so low if, in fact, they are highly polluted? Although the final answer here awaits further work, present data suggest that lowland basins should manifest lower cationic denudation compared to highland basins under natural conditions (Reynolds and Johnson, 1972). In highly industrialized regions cationic denudation is probably more an artifact of human activities than it is nature's. In New England, then, such small upland streams as Hubbard Brook more accurately reflect the natural state of cationic denudation of the landscape than do larger streams.

Our long-term estimates of cationic denudation at Hubbard Brook, then, allow us to estimate the relative importance of external and internal sources of H^+ ions as weathering agents. Our evaluation of cationic denudation is 2.0×10^3 Eq/ha·yr, which should be balanced by the sum of the net external and net internal supply of H^+ ions (Figure 40). The external supply rate is $\sim 1.0 \times 10^3$ Eq/ha·yr (Table 12) and by difference the internal source becomes $\sim 1.0 \times 10^3$ Eq/ha·yr. This suggests that under prevailing biologic and chemical conditions external and internal generation of H^+ ions play about equal roles in driving the weathering reaction at Hubbard Brook. This conclusion is based on the assumption that all H^+ ions entering the ecosystem from both external and internal sources are consumed in the weathering reactions.

Weathering Mode

Cationic denudation (net export plus biomass accumulation) values accurately portray the quantitative dimensions of chemical weathering but do not disclose any information regarding the qualitative aspects of chemical weathering. For some insight into the latter problem at Hubbard Brook we can look at the relative rates at which various rock-derived elements are weathered. The chemical weathering for elemental Ca, Na, Mg, K, Si, and Al at Hubbard Brook, 21.1, 5.8, 3.5, 7.1, 18.1, and 1.9 kg/ha·yr, respectively, are in serial order Ca > Si > K > Na > Mg > Al (Table 20). This sequence is distinctly different from that predicted by Polynov (1937) for general granitic terrain and that found by Anderson and Hawkes (1958) for New England in particular. It is significant that if the biomass accretion component is neglected, the relative order changes substantially. That is, the net loss of elemental Ca, Na, Mg, K, Si, and Al in dissolved load only is 11.6, 5.6, 2.6, 1.0, 18.0, and 1.9 kg/ha·yr,

TABLE 20. Differential Chemical Weathering at Hubbard Brook.

Element	Abundance in bedrock[a], %	A Annual release from bedrock by weathering,[b] kg/ha	B Amount contained in 1,500 kg of bedrock, kg	Differential weathering ratio[c] = (**A/B**) × 100, %
Ca^{2+}	1.4	21.1	21.1	100
Na^{+}	1.6	5.8	24.1	24
Mg^{2+}	1.1	3.5	16.5	21
K^{+}	2.9	7.1	43.6	16
Al^{3+}	8.3	1.9	124.8	2
Si^{4+}	30.7	18.1[d]	461.7	2

[a]Taken from Johnson et al. (1968).

[b]Based on net output of dissolved substances (Table 11) plus living and dead biomass accumulation, which for Ca^{2+}, Na^{+}, Mg^{2+}, and K^{+} is 9.5, 0.17, 0.9, and 6.1 kg/ha·yr, respectively.

[c]Normalized to calcium, i.e., assuming the complete extraction of calcium from 1,500 kg of bedrock.

[d]Assuming that dissolved silica is in the form of SiO_2 (Table 11), and 0.1 kg Si per hectare is exported in stream water as organic particulate matter.

respectively (Table 12). This results in a serial order of Si > Ca > Na > Mg > Al > K (but see Table 19), which is still at considerable variance with other published elemental mobility series. These disparities suggest that differential mobility series have limited value as a general index to chemical weathering activity, until such time as collection, sampling, and analytic techniques are normalized to some common basis. This long-term, comprehensive data collected in the Hubbard Brook Ecosystem Study allow us to compensate for the effects of precipitation chemistry, biologic activity, hydrologic dilution, and seasonal vaiation in characterizing stream-water chemistry. To the extent that these factors are crucial in establishing a relative mobility series, ignoring them must produce highly unreliable results. Our experience suggests that the most important factor in this regard is the growth status of the forest, which may mask the actual rate at which rocks are chemically decomposing within the system.

Because of our detailed cationic budget data, however, we can make other, more direct inferences about differential chemical weathering at Hubbard Brook. The absolute cationic effluxes we have observed, when compared to the absolute abundances of these cations in the bedrock at Hubbard Brook, testify that differential chemical weathering is actively underway (see Table 20). The calcium minerals of the bedrock are obviously the most prone to weathering. Some 1.5 metric tons per hectare of bedrock must undergo weathering every year to provide the requisite calcium to the system. If we assume that this calcium-based value (1.5 tons of rock per hectare year) is the lower limit for the rate of rock weathering, we can then calculate the fraction of each element extracted by the weathering process (Table 20). These latter estimates seem to be compatible with what is known about the vulnerability of various silicate minerals to chemical attack (Goldich, 1938; Marshall, 1964); i.e., calcium-rich minerals are the least stable and potassium-rich minerals are the most stable in an acidic chemical regime.

One final point: When compared with the usual published chemical analyses for major streams in New England (Johnson et al., 1972), the headwater streams of the HBEF appear to be rather unusual in chemical makeup; i.e., they are dominated by strong acids (sulfuric and nitric). Because of its physiographic location, the HBEF (Figure 2) represents the incipient stages of drainage water; i.e., precipitation has but recently percolated through the soil zone and is appearing in variable source areas or first-order stream channels. These acids, however, are in the process of being neutralized by chemical weathering reactions. The relatively low pH typical for HBEF stream water (pH 4–5) precludes the presence of significant quantities of ionized carbonic acid (Stumm and Morgan, 1970). Significantly, the SO_4^{2-} content of New England stream water is fixed at this stage at 4–6 mg/l and it stays essentially at this concentration during the remainder of its transit to the sea (Johnson et al., 1972; Pearson and Fisher, 1971). As the water moves downstream it is progressively neutralized by weathering reactions. As the pH rises above 5, carbonic acid can dissociate and carbonation weathering reactions may proceed. The rate of neutralization varies somewhat as a function of local streamflow rate and local bedrock chemistry but our experience at Hubbard Brook suggests that the major chemical composition is determined within the first few hundred meters of streamflow. Bicarbonate alkalinity predominates downstream as shown by chemical analyses in the mainstem of Hubbard Brook itself.

Addendum

The release of ionic substances from primary minerals by weathering is an important source of nutrients in soils and soil solutions of forest ecosystems (e.g., Likens et al., 1967; Mast and Drever, 1990; Williams et al., 1986; Yuretich and Batchelder, 1988). Initially, we estimated rates of weathering as the long-term cationic denudation rate. A major insight relative to this estimate was that long-term sequestration of cations by the biomass must be included in the calculation. The role of cation exchange and secondary mineral formation was identified, but could not be quantified or differentiated. The full, mass balance for this calculation of cationic denudation on an annual basis is:

$$P_i + W_i = S_i + \Delta B_i \pm \Delta O_i \pm \Delta X_i \pm \Delta M_i \qquad (1)$$

where P_i = atmospheric input of element i, W is the weathering release from primary minerals, S is dissolved loss in stream water, ΔB is net storage in biomass, ΔO is net storage in the soil organic matter pool, ΔX is the net change in the exchangeable pool, and ΔM is the net change in the secondary mineral pool. Since direct estimates of W_i, ΔX_i, ΔM_i, and ΔO_i currently are quantitatively unknown or poorly known at Hubbard Brook, it has been proposed that "net soil release" is a more appropriate term than "weathering" or "cationic denudation" (Likens et al., 1994) and is calculated as:

$$\text{Net soil release} = W_i \pm M_i \pm \Delta X_i \pm \Delta O_i = S_i + \Delta B_i - P_i \qquad (2)$$

Recent data on the biogeochemistry of potassium (Likens et al., 1994) suggest that secondary minerals (particularly vermiculite) and the cation exchange complex probably play important roles in regulating the loss of K in stream water on an annual basis at Hubbard Brook.

Bailey (1994) used a detailed analysis of the isotopic composition of strontium, coupled with a mass balance analysis of both calcium and strontium, in an attempt to distinguish between weathering release and net exchange with various ecosystem pools (e.g., cation exchange complex, forest floor). This work was done in the Cone Pond watershed, some 7 km from the HBEF. He estimated the weathering rate for calcium to be only about 30% of the value estimated by a standard mass balance.

6
Nutrient Cycles

Chemical elements without a prominent gaseous phase at normal biologic temperature, such as calcium, magnesium, and potassium, have what is referred to as a sedimentary biogeochemical cycle (Odum, 1959). That is, flux and cycling are affected primarily by hydrologic factors (including dissolution, erosion, and sedimentation), landslides, vulcanism, and biologic agents. The presence of a prominent gaseous component in the biogeochemical cycle (e.g., C, N, and S) greatly complicates measurement and analysis, as discussed previously.

The Calcium Cycle

To illustrate a sedimentary cycle, we shall summarize the average (11 yr) biogeochemical relationships for calcium at Hubbard Brook (Figure 41). Bulk precipitation contributes 2.2 kg/ha of calcium to the ecosystem each year. In addition, some 21.1 kg of Ca per hectare·year are released within the ecosystem by weathering. Because this 55-yr-old forest is accumulating biomass, 5.4 kg of Ca per hectare are stored above ground and 2.7 kg/ha·yr below ground in the annual vegetation growth increment. There is an additional net accumulation of 1.4 kg/ha·yr in the dead biomass of the forest floor. The remainder, 13.9 kg/ha·yr, is lost from the system in drainage water. Most (98.6%) of this loss is in the dissolved form. The annual net loss from the ecosystem (11.7 kg/ha) is only 2% of the total available calcium (510 kg/ha) within the ecosystem and about 19% of the amount circulated annually (uptake = 62.2 kg/ha) by the vegetation.

Most of the calcium in the ecosystem, more than 99% is present in the soil complex, whereas only about 0.5% (484 kg/ha) is bound in the vegetation. Nevertheless, about 12% of the available nutrient pool is cycled (taken up) by the vegetation each year. Meteorologic input accounts for only about 3.5% of all calcium taken up by vegetation annually, whereas weathering generates about 34% of the annual vegetation uptake. Of the annual amount of calcium taken up by vegetation,

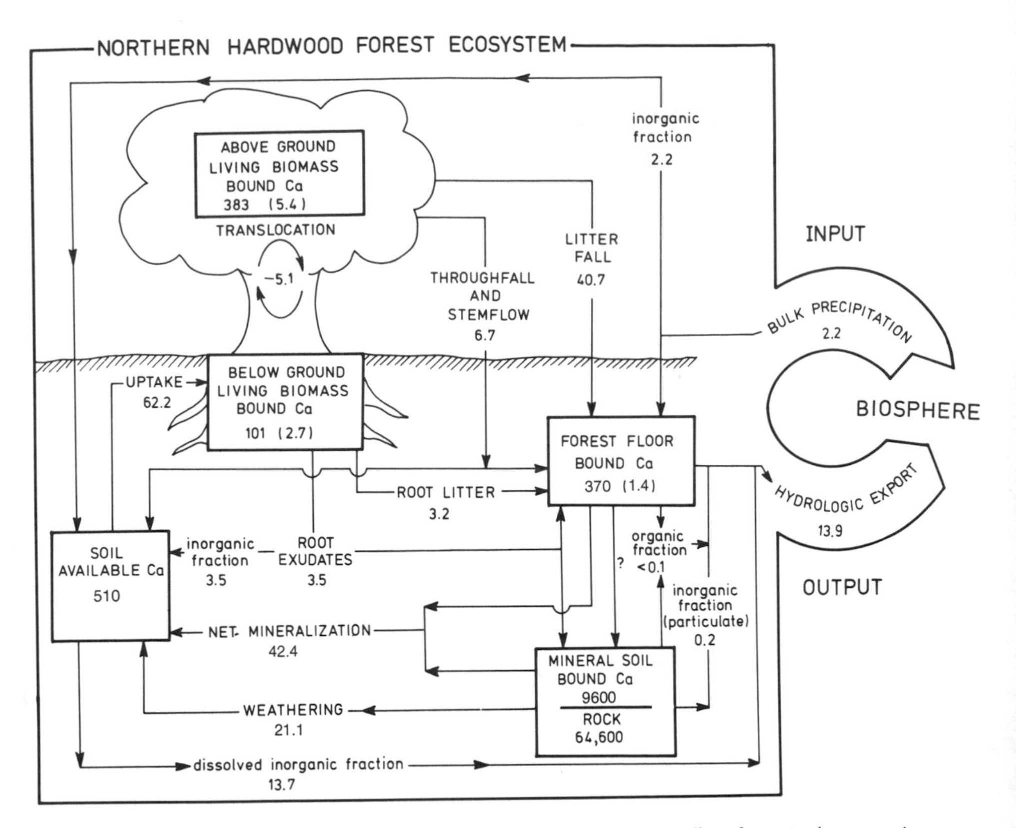

FIGURE 41. Average annual calcium budget for an aggrading forested ecosystem at Hubbard Brook during 1963–1974. Standing crop values are in kg/ha and calcium fluxes are in kg/ha·yr. Values in parentheses represent annual accretion rates.

FIGURES 41 and 42. Ecosystem Diagrams.

General. The boxes within the diagram represent standing stocks or pools of elements. The values within these boxes are the size of the elemental pool and are in kg/ha. Because the Hubbard Brook Forest is an aggrading ecosystem with an annual net accumulation of biomass, the rate at which the element is accumulating each year is given in parentheses within the appropriate box in kg/ha·yr. The arrows between the various pools represent annual fluxes and are given in kg/ha·yr.

The last cutting of the forest at Hubbard Brook took place between 1909 and 1917. Indications are that the majority of the cutting took place closer to 1917 than to 1909. Therefore, we are making the assumption that forest regrowth was initiated around 1915. The majority of the measurements, which form the basis for the diagrams, were made around 1970. Where the age of the forest was needed (e.g., in the calculation of forest floor accretion) 55 yr was used.

Pools. *Aboveground living biomass:* Living biomass was measured by Whittaker et al. (1974). At the same time representative samples were taken and analyzed by Likens and Bormann (1970) for various elemental constituents.

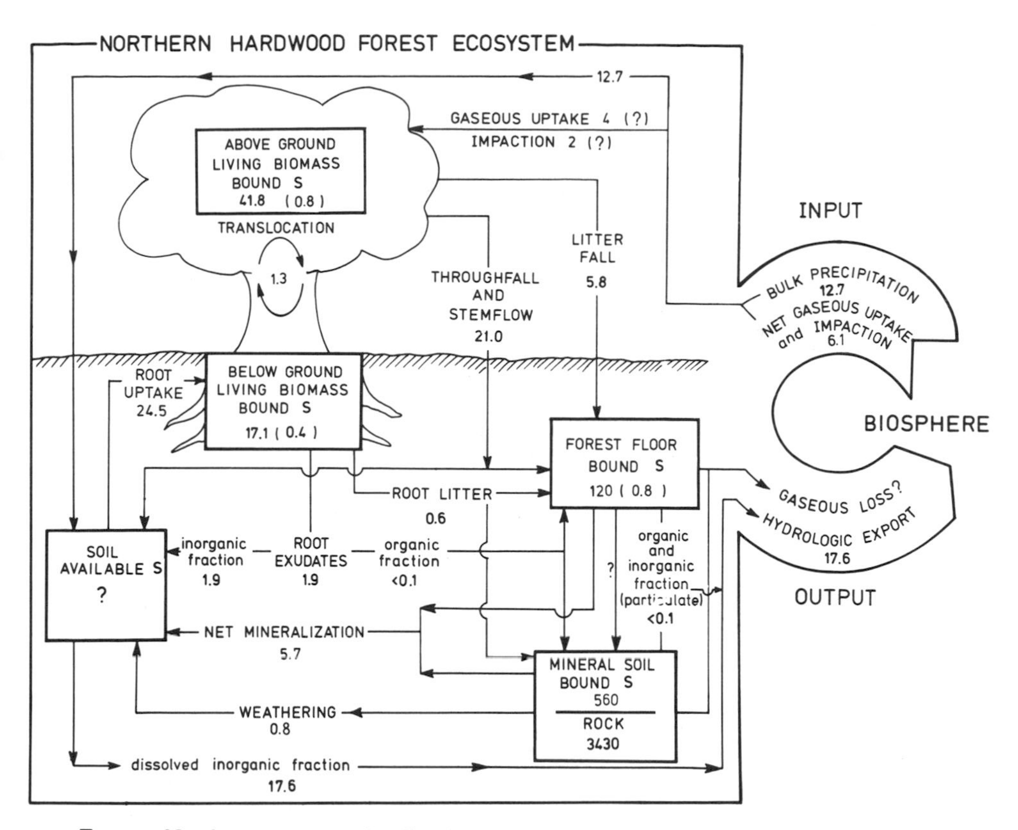

FIGURE 42. Average annual sulfur budget for an aggrading forested ecosystem at Hubbard Brook during 1963–1974. Standing crop values are in kg/ha and sulfur fluxes are in kg/ha·yr. Values in parentheses represent annual accretion rates.

Whittaker et al. (1974) also measured the productivity of the forest for two pentads (1956–1960 and 1961–1965). The biomass accretion values in the diagrams represent a mean of these two pentads and are multiplied by the elemental concentrations of Likens and Bormann (1970).

Belowground living biomass: Belowground biomass was measured and elemental composition determined in the same manner as the aboveground living biomass (Whittaker et al., 1974; Likens and Bormann, 1970).

Forest floor: The biomass and the elemental composition of the upper organic layers of the soil (to an average depth of 8.9 cm) was measured by Gosz et al. (1976) and Dominski (1971). The biomass of larger litter (e.g., limbs and boles of trees) on the forest floor was measured by Covington (1976) and its elemental composition was determined from Gosz et al. (1972).

The rate of forest floor accretion was estimated by Covington (1976) by measuring the biomass of a successional sequence of northern hardwood stands ranging in age from 3 to 200 yr. Elemental composition is from Gosz et al. (1976).

In some northern hardwood ecosystems, biomass accretion occurs in the form of dead wood which is not part of the living biomass and not yet part of the forest

FIGURES 41 and 42. Continued.
floor. Input–output–standing crop relationships of the dead wood pool at Hubbard Brook are under study and data suggest that dead wood in the 55-yr ecosystem is in steady state where input approximately equals output.

Mineral soil and rock: The amount of organic and inorganic matter in the mineral soil and of rock in the soil profile to a depth of 45 cm was measured by Dominski (1971). The elemental composition of mineral soil is from Gosz et al. (1976), Lunt (1932), and Johnson et al. (1968); the elemental composition of rock is from Johnson et al. (1968) assuming 50% unweathered Kinsman quartz monzonite and 50% unweathered Littleton gneiss. Although an arbitrary soil depth of 45 cm was used, we believe the majority of the active soil processes (biogeochemical) occur within the upper 45 cm of soil. During the period of this study we assume that changes in the chemical content of the mineral soil are very small and do not significantly affect the budget calculations.

Available soil nutrients: Estimated from soil depth, bulk density, and elemental analyses (Lunt, 1932).

Elemental Fluxes. *Bulk-precipitation:* Chemical inputs in rain and snow to the ecosystem have been measured since 1963. Numbers used here are weighted mean annual inputs (see Table 11).

Net gaseous uptake and impaction of aerosols: This was calculated as net uptake = Losses (net biomass accumulation + net forest floor accumulation + export of dissolved substances and particulate matter) − gains (weathering + precipitation input).

Litterfall: The fall of litter from forest vegetation to the forest floor was measured and its elemental composition, determined by Gosz et al. (1972).

Translocation: The movement of chemicals back into woody tissues at the time of autumn senescence for leaves, is calculated as translocation = (production + bud litter + root exudates + throughfall and stemflow) − (uptake − shrub and herb litter).

Throughfall and stemflow: The washing and leaching of chemicals from the forest canopy by precipitation was determined by Eaton et al. (1973).

Root exudation: The loss of chemicals from roots by root exudates was measured by Smith (1976).

Root litter: The biomass of root litter was estimated by Whittaker et al. (1974). Elemental composition of roots was taken from Likens and Bormann (1970).

Net mineralization: This transformation of organically bound element to inorganic form assumes that there is no microbial immobilization of the element. Net mineralization = (annual uptake by plants + annual hydrologic export of inorganic element) − (annual precipitation input of inorganic element + annual transfer of inorganic element in root exudates + annual transfer of inorganic element in throughfall and stemflow + weathering).

Uptake: Uptake by forest vegetation was calculated as uptake = litterfall + net throughfall and stemflow + root exudates + root litter + living biomass accumulation.

Hydrologic export: Export of dissolved chemicals represent means for 1963–1974 (Table 11). Particulate losses represent means for 4 yr (Table 10).

Weathering: Calculated on the basis of 1,504 kg of rock weathered each year. Elemental composition of the rock substrate is given by Johnson et al. (1968). See text for further details.

litter, stemflow, throughfall, and root exudates return some 87% to the forest floor; 13% of the uptake is accumulated by the forest vegetation. Interestingly, the living vegetation releases more calcium in root exudates than is lost as root litter. Gosz et al. (1973) calculated that some 17.3 kg of Ca per hectare were released by decomposition and leaching of current litter each year at Hubbard Brook. Net mineralization from the entire soil mass is estimated at 42.4 kg of Ca per hectare (Figure 41). Of the five inputs to the available nutrient compartment, net mineralization is the largest (56% of the total), followed by weathering (28%), leaching from the forest canopy (9%), root exudates (5%), and precipitation (3%). Although there are large quantities of calcium made available and circulated within the intrasystem cycle of the ecosystem, the growing forest has minimal losses. That is, the forest ecosystem is very efficient in cycling and retaining calcium.

Some of the values in Figure 41 differ from those published previously (Likens and Bormann, 1972). This primarily is the result of new or additional data for the living and dead biomass components of the HBEF and also because the soil depth used here is only 45 cm instead of 61 cm as used previously.

The Sulfur Cycle

Such elements as carbon, sulfur, and nitrogen have a prominent gaseous phase at normal biologic temperatures, and this greatly complicates quantitative evaluations of the biogeochemical flux. To provide an example of this type of biogeochemical relationship, we shall use sulfur (Figure 42).

The biogeochemistry of sulfur plays a key role in the function of forested ecosystems at Hubbard Brook. Sulfate is the dominant anion in both precipitation and stream water. Approximately 99% of the sulfur in the ecosystem at any one moment is found in the soil complex and the remaining 1% occurs in the living biomass. The vegetation takes up sulfur at a rate of 28.5 kg/ha·yr and accretes it at a rate of 1.2 kg/ha·yr. Weathering generates an amount of sulfur equivalent to about 3% of all sulfur taken up annually by vegetation and meteorologic input accounts for approximately 66% of the annual vegetation uptake (Figure 42). The sulfur budget of the forest ecosystems at Hubbard Brook is therefore dominated by meteorologic inputs.

A summation of the values for biomass accretion, weathering, precipitation input, and stream-water output shows a sulfur inbalance of 6.1 kg/ha·yr, suggesting some additional source(s) of sulfur for the ecosystem. We believe that dry deposition of sulfur may provide the necessary net sulfur input of 6.1 kg/ha·yr (Figure 42). It is well known that plants can utilize SO_2 directly from the atmosphere (Cowling et al.,

1973; Hill, 1971; Hoeft et al., 1972), and various authors have suggested input from aerosols and dust impacted on vegetation surfaces (Carlisle et al., 1967; Duvigneaud and Denaeyer-DeSmet, 1964; Eriksson, 1952; Tamm and Troedsson, 1955; White and Turner, 1970). Our throughfall and stemflow data (Table 7) suggest that sulfurous aerosols may be impacted on vegetation surfaces in large amounts during the summer. We do not know the actual proportions of these two meteorologic sources of sulfur at Hubbard Brook but we estimate that the net gaseous input is about double that from aerosol deposition on an annual basis (Eaton et al., 1976). The combination of net gaseous uptake and aerosol deposition account for about 32% of the total meteorologic input or 35% of the annual hydrologic export. Obviously such inputs must be carefully assessed in biogeochemical studies.

In contrast to the calcium cycle, the ecosystem is not nearly as efficient in retaining sulfur. Some 94% of the sulfur added to the ecosystem annually is lost in stream water.

Quantitative measures of gaseous flux, microbial transformations, and soil standing stocks have not been fully evaluated. These are some of the objectives of ongoing studies at Hubbard Brook.

Nutrient Budget Relationships at Hubbard Brook

The allocation of budgetary items for the major nutrients at Hubbard Brook is given in Tables 21 and 22. These values, along with those of Tables 11–13, suggest several important conclusions regarding the biogeochemistry of the northern hardwood ecosystem. These are: (a) In relation to gross losses, nutrient inputs in bulk precipitation represent a significant addition to the ecosystem. Such inputs can be the major source of nutrients in terrestrial ecosystems low in weathering substrates (e.g., Art et al., 1974). (b) For some elements, such as N, S, and Cl, dry deposition or biologic gaseous fixation can provide significant input to forest ecosystems. (c) Based on these budgets, the forest ecosystem shows absolute gains in C, N, S, P, and Cl and absolute losses in Si, Ca, Na, Al, Mg, and K. Losses of the latter substances from the intrasystem nutrient cycle are made up by weathering primary minerals (Figure 1) and the losses therefore imply a decrease in weathering substrate. (d) Weathering is the major source of calcium, potassium, magnesium, and sodium in the ecosystem. (e) Assuming that the mineral horizons of the soil are near steady state, a large percentage (greater than 40%) of potassium, nitrogen and calcium input from meteorologic sources and released by weathering is stored annually by living and dead biomass within the ecosystem. The forest floor is a particularly effective storage site for nitrogen. In contrast, only about 2% of the sodium and 10% of the sulfur added to the ecosystem is stored in the biomass, and the remainder, more than 90%, is

TABLE 21. Standing Stocks and Annual Biogeochemical Fluxes for a 55-yr old Forested Ecosystem at Hubbard Brook.

Component	Chemical Element							
	Ca	Mg	Na	K	N	S	P	Cl
	Standing stock (kg/ha)							
Aboveground biomass	383	36	1.6	155	351	42	34	*
Belowground biomass	101	13	3.8	63	181	17	53	*
Forest floor	372	38	3.6	66	1256	124	78	*
	Annual flux (kg/ha·yr)							
Bulk precipitation input	2.2	0.6	1.6	0.9	6.5	12.7	0.04	6.2
Gaseous or aerosol input	*	*	*	*	14.2	6.1	*	?
Weathering release	21.1	3.5	5.8	7.1	0	0.8	?	*
Streamwater output								
Dissolved substances	13.7	3.1	7.2	1.9	3.9	17.6	0.01	4.6
Particulate matter	0.2	0.2	0.2	0.5	0.1	<0.1	0.01	*
Vegetation uptake	62.2	9.3	34.8	64.3	79.6[a]	24.5[a]	8.9	*
Litter fall	40.7	5.9	0.1	18.3	54.2	5.8	4.0	*
Root litter	3.2	0.5	0.01	2.1	6.2	0.6	1.7	*
Throughfall and stemflow	6.7	2.0	0.3	30.1	9.3	21.0	0.7	4.4
Root exudates	3.5	0.2	34.2	8.0	0.9	1.9	0.2	1.8
Net mineralization	42.4	6.1	0.1	20.1	69.6	5.7	?	?
Aboveground biomass accretion	5.4	0.4	0.03	4.3	4.8	0.8	0.9	*
Belowground biomass accretion	2.7	0.3	0.12	1.5	4.2	0.4	1.4	*
Forest floor accretion	1.4	0.2	0.02	0.3	7.7	0.8	0.5	*

*small, unmeasured

[a]root uptake

TABLE 22. Allocation of Budgetary Items in Percent for Watershed Ecosystems of the Hubbard Brook Experimental Forest.

	Ca	K	Mg	Na	N	S
Source						
Precipitation input	9	11	15	22	31	65
Net gas or aerosol input	—	—	—	—	69	31
Weathering release	91	89	85	78	—	4
Storage or loss						
Biomass accumulation						
Vegetation	35	68	17	2	43	6
Forest floor	6	4	5	<1	37	4
Streamflow						
Dissolved substances	59	22	74	95	19	90
Particulate matter	<1	6	5	3	<1	<1

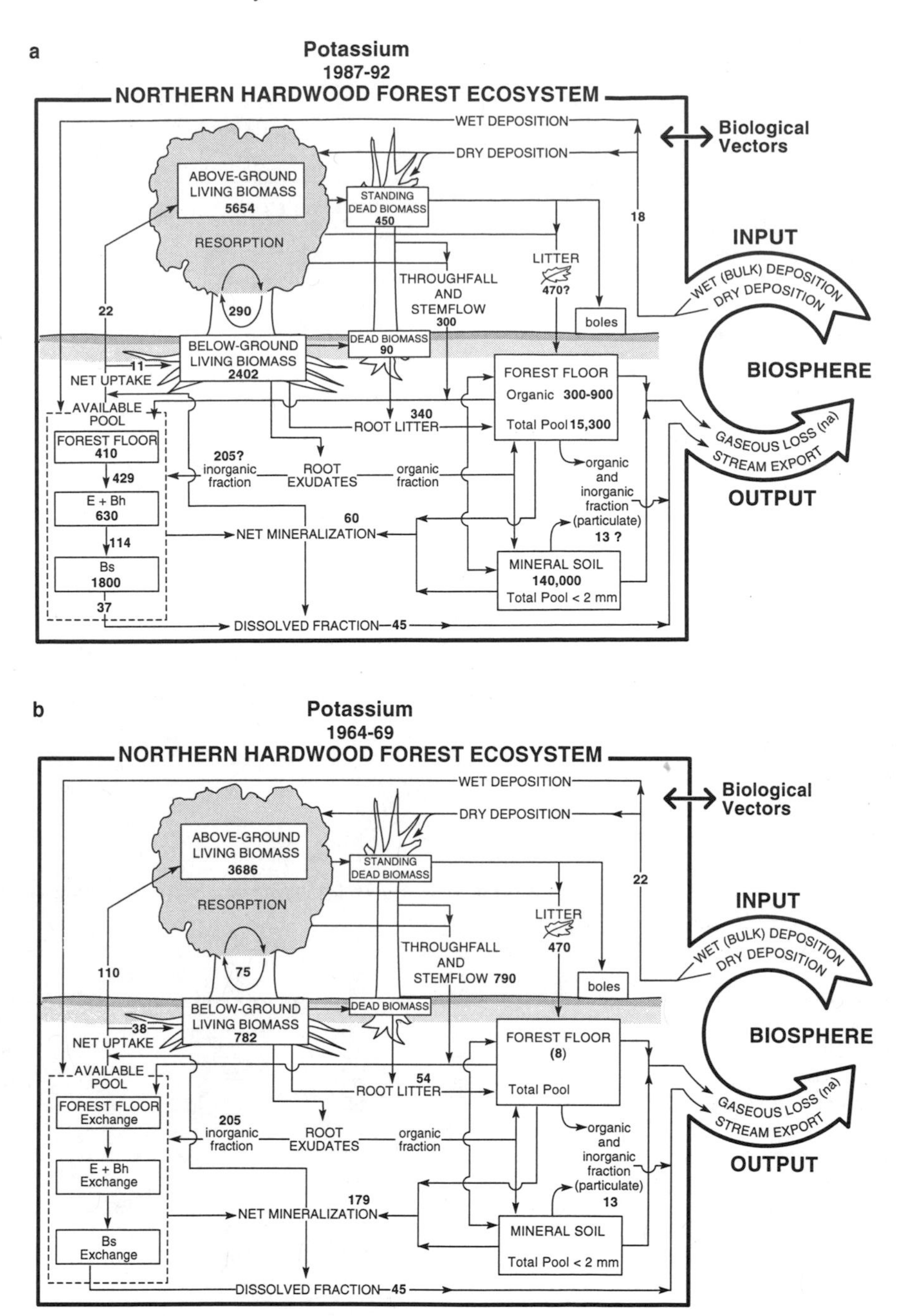

FIGURE 43. Ecosystem pools (boxes) and fluxes (arrows) of K during 1964–1969 (**a**) and 1987–1992 (**b**) for Watershed 6 of the HBEF. Values in mol/ha or mol/ha·yr. From Likens et al., 1994. Reprinted by permission of Kluwer Academic Publishers.

lost in streamflow. (f) Root exudates are very important to the intrasystem cycle of sodium, whereas stemflow and throughfall are particularly important for cycling sulfur and potassium, and litter plays a major role in the cycling of phosphorus, nitrogen, magnesium and calcium.

Addendum

A series of detailed monographs on the biogeochemistry of major chemical elements at Hubbard Brook is planned. The first of these, on potassium, has been published (Likens et al., 1994). Because of changes in the long-term chemistry of precipitation and particularly because of the decline in the growth of the forest vegetation at Hubbard Brook, significant changes in the flux and cycling of potassium occurred during the early (1964–1969) period of study and the current (1987–1992) period. For example, storage of K in biomass, net soil release and throughfall all decreased, whereas resorption increased from the earlier pentad to the later one (Figure 43). Much more detailed information on the chemistry of soils, soil solutions and roots is available now than in 1977 (e.g., Driscoll et al., 1985; Nodvin et al., 1988; Johnson, 1989; Huntington et al., 1989; Fahey and Hughes, 1994; Likens et al., 1994).

7 The Northern Hardwood Ecosystem at Hubbard Brook in Relation to Other Forest Ecosystems in the World

Forests and woodlands cover some $57 \times 10^6 \text{km}^2$, which is about 38% of the total continental area or 11% of the Earth's surface. Despite this relatively small area, 92% of the Earth's plant biomass and 46% of its annual net primary production come from forest (Table 23). The 79.9 billion metric tons of dry plant matter produced (net) each year by forests exceeds the total net primary production of the oceans, even though the oceans are more than six times larger in area. The animal biomass in these forests represents 68% of the total found on continental land masses (Whittaker and Likens, 1973).

The structure and well-being of forest ecosystems to a large extent reflect the chemical balance that is extant for the ecosystem. It is of interest to examine briefly how the biogeochemical data from the Hubbard Brook Ecosystem Study relate to some other studies done at various places in the world. Budgets of precipitation input/stream-water output for a variety of forest ecosystems of the world are given in Table 24. These data describe the general nutrient flux for these systems.

Data in Table 24 encompass a wide range in ecosystem analysis and sophistication. In some cases these data represent averages of several watersheds for several years. In others, data are scanty and are derived from a poorly defined hydrologic system. These data, therefore, although very interesting and informative, should be treated with some caution. Moreover, these relationships may become modified as more information on net exchange of gases and dry deposition of aerosols becomes available. We have not incorporated adjustments for net gaseous input and dry deposition of aerosols in Table 24, except in the case of N and S for Hubbard Brook, for which data have been obtained from an analysis of the total ecosystem. Considerable effort should be devoted to the problem of measuring dry deposition and gaseous exchange in forested ecosystems in the future.

In general, the findings from these studies show a common pattern with those of the Hubbard Brook Ecosystem Study. Several conclusions emerge: (1) Chemicals in precipitation are significant additions to the

TABLE 23. Estimates of Area, Biomass, and Productivity for Major Components of the Biosphere.[a]

	Area, 10^6 km^2	Plant biomass, 10^9 dry tons	Net plant productivity, 10^9 dry tons/yr	Animal biomass, 10^6 dry tons	Animal productivity, 10^6 dry tons/yr
Forest and woodlands	57	1,700	79.9	680	442
Total continental area	149	1,837	117.5	1,015	827
Total ocean	361	4	55.0	998	3,057
Total biosphere	510	1,841	172.5	2,013	3,884

[a]It is assumed that dry matter × 0.45 = carbon. Modified from Whittaker and Likens (1973, 1975).

budgets for a variety of forest ecosystems. (2) Precipitation inputs of inorganic nitrogen and phosphorus exceed losses in stream water for forest ecosystems. There are three exceptions to the generalization that phosphorus inputs in precipitation exceed outputs. Although most of the watersheds in the Western Cascades Range of Oregon showed outputs larger than inputs during the years studied, some watersheds showed net gains (Fredriksen, 1975) and the pattern was therefore not clear. The other exceptions, Finland and Taughannock Creek, NY, include drainage from agricultural land. Human activities or production of wastes can readily overwhelm the capacity of natural ecosystems to conserve phosphorus within the ecosystem (Likens and Bormann, 1974a). (3) In general, there are net losses of calcium, magnesium, sodium, and potassium from forest ecosystems that must be made up by weathering. (4) Geologic substrates play a major role in determining the amount and composition of chemicals lost in stream water. Watersheds on relatively soluble bedrock (England; Taughannock Creek, NY; Walker Branch, TE; El Verde, Puerto Rico; and Carnation Creek, Vancouver Island) lose large quantities of the major cations via streamflow. (5) Forest ecosystems in coastal locations or otherwise dominated by maritime air masses have relatively large inputs of sea salts in precipitation (Art, 1976). (6) A close correlation between the amount of precipitation and the loss of nutrients may apply to individual watersheds, such as the HBEF, but as a general principle there is no close correlation between the amount of precipitation and loss of nutrients in forest ecosystems with different climates. (7) Net loss or gain values for sulfur budgets are significantly affected by anthropogenic contributions to precipitation. Contaminated air masses over Hubbard Brook; Taughannock Creek, NY; Walker Branch, TE; Birkenes, Norway; and, to a lesser extent, southern Sweden, the southeastern United States, and Vancouver Island, Canada, produce high levels of sulfur in local precipitation. (8) Forest ecosystems in a variety of

TABLE 24. Nutrient Budgets for Various Terrestrial Ecosystems of the World (kg/ha·yr).

	Location[a]	Precipitation input	Stream-water output	Net gain or loss	Annual precipitation, cm	Dominant vegetation[b]	Geology[c]
			Calcium				
Temperate: mostly angiosperm and deciduous forest							
	Coweeta, NC, U.S. (1)	6.2	6.9	−0.7	185	Q, C, Ac	M_g, M_{sh}
	Hubbard Brook, U.S. (2)	2.2	13.9	−11.7	130	Ac, F, B	I_g, M_{sh}
	Long Island, NY, U.S. (3)	3.3	9.6[d]	−6.3	124	Q, Pin	S_s
	Pago Catchment, Australia (4)	7.0	7.1	−0.1	150	E	M_{sh}, I_g
	S.E. U.S. (5)	6	19	−13	127	Q, Pin	S_s, S_{sh}, M_{sh}
	Silverstream, New Zealand (6)	6	26	−20	135	N	S_s, S_{sh}, M_s, M_{sh}
	Taughannock Creek, NY, U.S. (7)	14.9	182	−167	96	Ac, Ti, Ts	S_{sh}, S_c
	Tesuque Watersheds, NM, U.S. (8)	6.8	13.1	−6.3	72	Po	I_g, M_g
	Walker Branch, TE, U.S. (9)	14.3	148	−134	155	Q, C	S_c
Temperate: mostly coniferous and evergreen forest							
	Birkenes Watershed, Norway (10)	2.7	14.1	−11.4	134	B, Pin, Pic	I_g
	Blue Range Catchment, Australia (4)	6.2	5.5	+0.7	132	Pin	I_b, M_{sh}
	Carnation Creek, Vancouver Island, Canada (11)	3.7	57.7	−54	315	Ab, Ts, Th	M_b, M_s, M_c
	Cedar River, WA, U.S. (12)	2.8	4.5[d]	−1.7	136	Ps	S_s
	ELA, Ontario, Canada (13)	3.8	6.0	−2.2	83	Pin, Pic	I_g, M_g
	Finland (14)	2	12	−10	57	Pin, Pic, B	I_g, M_g
	Storsjön, Sweden (15)	3.6	12.2	−8.6	87	Pin, Pic	I_g
	Tesuque Watersheds, NM, U.S. (8)	6.4	10.1	−3.7	74	Pic, Ab, Pin	I_g, M_g
	Tesuque Watersheds, NM, U.S. (8)	6.7	22.3	−15.6	48	Pin, J	I_g, M_g
	Velen, Sweden (16)	5.9[e]	11.1[e]	−5.2	72	Pin, Pic	M_g, I_g
Temperate: bog vegetation							
	Rough Sike Catchment, England (17)	9.0	53.8[f]	−44.8	213	Bog	S_c, S_s, S_{sh}
Tropical: angiosperm, mostly evergreen forest							
	El Verde, Puerto Rico (18)	21.8	43.1	−21.3	376	Trp	I_b, S_{sh}, S_s
	Rio Negro, Brazil (19)	0.8	4.7[g]	−3.9	200	Trp	S_{sh}, M_{sh}, M_g

			Magnesium				
Temperate: mostly angiosperm and deciduous forest							
	Coweeta, NC, U.S.	1.3	3.1	−1.8	185	Q, C, Ac	M_g, M_{sh}
	Hubbard Brook, U.S.	0.6	3.3	−2.7	130	Ac, F, B	I_g, M_{sh}
	Long Island, NY, U.S.	2.1	7.3[d]	−5.2	124	Q, Pin	S_s
	Pago Catchment, Australia	1.5	4.5	−3.0	150	E	M_{sh}, I_g
	S.E. U.S.	2	6	−4	127	Q, Pin	S_s, S_{sh}, M_{sh}
	Silverstream, New Zealand	9	13	−4	135	N	S_s, S_{sh}, M_s, M_{sh}
	Taughannock Creek, NY, U.S.	1.4	34.8	−33.4	96	Ac, Ti, Ts	S_{sh}, S_c
	Tesuque Watersheds, NM, U.S.	0.7	4.2	−3.5	72	Po	I_g, M_g
	Walker Branch, TE, U.S.	2.1	77.1	−75.0	155	Q, C	S_c
Temperate: mostly coniferous and evergreen forest							
	Birkenes Watershed, Norway	2.3	5.0	−2.7	134	B, Pin, Pic	I_g
	Blue Range Catchment, Australia	1.3	8.7	−7.4	132	Pin	I_g, M_{sh}
	Carnation Creek, Vancouver Island	3.0	10.4	−7.4	315	Ab, Ts, Th	M_b, M_s, M_c
	ELA, Ontario, Canada	0.9	2.4	−1.5	83	Pin, Pic	I_g, M_g
	Finland	1	4	−3	57	Pin, Pic, B	I_g, M_g
	Storsjön, Sweden	1.9	8.7	−6.8	87	Pin, Pic	I_g
	Tesuque Watersheds, NM, U.S.	0.6	3.4	−2.8	74	Pic, Ab, Pin	I_g, M_g
	Tesuque Watersheds, NM. U.S.	0.7	10.8	−10.1	48	Pin, J	I_g, M_g
	Velen, Sweden	1.2[e]	3.2[e]	−2.0	72	Pin, Pic	M_g, I_g
Temperate: bog vegetation							
	Maesnant Catchment, mid-Wales[q] (20)	4.4	8.7	−4.3	164	Bog	S_{sh}, S_s
Tropical: angiosperm, mostly evergreen forest							
	El Verde, Puerto Rico	4.9	15.0	−10.1	376	Trp	I_b, S_{sh}, S_s
	Rio Negro, Brazil	2.0	3.0[g]	−1.0	200	Trp	S_{sh}, M_{sh}, M_g

(*continued*)

TABLE 24. Nutrient Budgets for Various Terrestrial Ecosystems of the World (kg/ha·yr) (*continued*)

	Location[a]	Precipitation input	Stream-water output	Net gain or loss	Annual precipitation, cm	Dominant vegetation[b]	Geology[c]
			Sodium				
Temperate: mostly angiosperm and deciduous forest							
	Coweeta, NC, U.S.	5.4	9.7	−4.3	185	Q, C, Ac	M_g, M_{sh}
	Hubbard Brook, U.S.	1.6	7.5	−5.9	130	Ac, F, B	I_g, M_{sh}
	Long Island, NY, U.S.	17	23[d]	−6	124	Q, Pin	S_s
	Pago Catchment, Australia	2.5	15.9	−13.4	150	E	M_{sh}, I_g
	S.E. U.S.	5	26	−21	127	Q, Pin	S_s, S_{sh}, M_{sh}
	Silverstream, New Zealand	50	62	−12	135	N	S_s, S_{sh}, M_s, M_{sh}
	Taughannock Creek, NY, U.S.	2.3	18.9	−16.6	96	Ac, Ti, Ts	S_{sh}, S_c
	Tesuque Watersheds, NM, U.S.	0.8	10.0	−9.2	72	Po	I_g, M_g
	Walker Branch, TE, U.S.	3.9	4.5	−0.6	155	Q, C	S_c
Temperate: mostly coniferous and evergreen forest							
	Birkenes Watershed, Norway	19.1[h]	30.3	−11.2	134	B, Pin, Pic	I_g
	Blue Range Catchment, Australia	2.2	16.2	−14.0	132	Pin	I_b, M_{sh}
	Carnation Creek, Vancouver Island, Canada	11.0	38.4	−27.4	315	Ab, Ts, Th	M_b, M_s, M_c
	ELA, Ontario, Canada	1.6	3.7	−2.1	83	Pin, Pic	I_g, M_g
	Finland	2	6	−4	57	Pin, Pic, B	I_g, M_g
	Storsjön, Sweden	12.9	37.5	−24.6	87	Pin, Pic	I_g
	Tesuque Watersheds, NM, U.S.	1.0	6.5	−5.5	74	Pic, Ab, Pin	I_g, M_g
	Tesuque Watersheds, NM, U.S.	0.7	4.9	−4.2	48	Pin, J	I_g, M_g
	Velen, Sweden	3.9[e]	9.2[e]	−5.3	72	Pin, Pic	M_g, I_g
Temperate: bog vegetation							
	Maesnant Catchment, mid-Wales	27.2	43.7	−16.5	164	Bog	S_{sh}, S_s
	Rough Sike Catchment, England	25.5	45.2[f]	−19.7	213	Bog	S_c, S_s, S_{sh}
Tropical: angiosperm, mostly evergreen forest							
	El Verde, Puerto Rico	57.2	64.5	−7.3	376	Trp	I_b, S_{sh}, S_s

			Potassium				
Temperate: mostly angiosperm and deciduous forest							
	Coweeta, NC, U.S.	3.2	5.2	−2.0	185	Q, C, Ac	M_g, M_{sh}
	Hubbard Brook, U.S.	0.9	2.4	−1.5	130	Ac, F, B	I_g, M_{sh}
	Long Island, NY, U.S.	2.4	3.9[d]	−1.5	124	Q, Pin	S_s
	S.E. U.S.	1	6	−5	127	Q, Pin	S_s, S_{sh}, M_{sh}
	Silverstream, New Zealand	5	13	−8	135	N	S_s, S_{sh}, M_s, M_{sh}
	Taughannock Creek, NY, U.S.	0.8	5.6	−4.8	96	Ac, Ti, Ts	S_{sh}, S_c
	Tesuque Watersheds, NM, U.S.	1.3	3.2	−1.9	72	Po	I_g, M_g
	Walker Branch, TE, U.S.	3.1	6.8	−3.7	155	Q, C	S_c
Temperate: mostly coniferous and evergreen forest							
	Birkenes Watershed, Norway	2.2	1.5	+0.7	134	B, Pin, Pic	I_g
	Carnation Creek, Vancouver Island, Canada	2.5	4.8	−2.3	315	Ab, Ts, Th	M_b, M_s, M_c
	Cedar River, WA, U.S.	0.8	1.0[d]	−0.2	136	Ps	S_s
	ELA, Ontario, Canada	1.1	1.2	−0.1	83	Pin, Pic	I_g, M_g
	Finland	2.4	4.6	−2[c]	57	Pin, Pic, B	I_g, M_g
	Storsjön, Sweden	1.7	4.0	−2.3	87	Pin, Pic	I_g
	Tesuque watersheds, NM, U.S.	1.4	2.1	−0.7	74	Pic, Ab, Pin	I_g, M_g
	Tesuque watersheds, NM, U.S.	1.1	1.2	−0.1	48	Pin, J	I_g, M_g
	Velen, Sweden	1.6[e]	2.5[e]	−0.9	72	Pin, Pic	M_g, I_g
Temperate: bog vegetation							
	Maesnant Catchment, mid-Wales	1.6	2.6	−1.0	164	Bog	S_{sh}, S_s
	Rough Sike Catchment, England	3.1	9.0[f]	−5.9	213	Bog	S_c, S_s, S_{sh}
Tropical: angiosperm, mostly evergreen forest							
	El Verde, Puerto Rico	18.2	20.8	−2.6	376	Trp	I_b, S_{sh}, S_s

(*continued*)

TABLE 24. Nutrient Budgets for Various Terrestrial Ecosystems of the World (kg/ha·yr) (*continued*)

	Location[a]	Precipitation input	Stream-water output	Net gain or loss	Annual precipitation, cm	Dominant vegetation[b]	Geology[c]
			Nitrogen				
Temperate: mostly angiosperm and deciduous forest							
	Coshocton, OH, U.S. (21)	20	2.5	+17.5	89	L, Pin	S_{sh}, S_c, S_s
	Hubbard Brook, U.S.	20.7[i,j]	4.0[i]	+16.7	130	Ac, F, B	I_g, M_{sh}
	S.E. U.S.	2[k]	1[k]	+1[k]	127	Q, Pin	S_s, S_{sh}, M_{sh}
	Silverstream, New Zealand	2.2	1.8	+0.4	135	N	S_s, S_{sh}, M_s, M_{sh}
	Taughannock Creek, NY, U.S.	9.7[i]	5.6[i]	+4.1	96	Ac, Ti, Ts	S_{sh}, S_c
	Walker Branch, TE, U.S.	8.7	1.8	+6.9	155	Q, C	S_c
Temperate: mostly coniferous and evergreen forest							
	Birkenes Watershed, Norway	14.5[i]	2.2[i]	+12.3	134	B, Pin, Pic	I_g
	Carnation Creek, Vancouver Island, Canada	2.7[i]	1.1[i]	+1.6	315	Ab, Ts, Th	M_b, M_s, M_c
	Cedar River, WA, U.S.	1.1	0.6[d]	+0.5	136	Ps	S_s
	ELA, Ontario, Canada	6.4	0.9	+5.5	83	Pin, Pic	I_g, M_g
	Finland	6	2	+4	57	Pin, Pic, B	I_g, M_g
	Storsjön, Sweden	10.0	2.3	+7.7	87	Pin, Pic	I_g
	Velen, Sweden	5.9[e,i]	0.4[e,i]	+5.5	72	Pin, Pic	M_g, I_g
	Western Cascades Range, OR, U.S. (22)	2.5[i]	1.2[i]	+1.3	200	Ps	I_b
Temperate: bog vegetation							
	Rough Sike Catchment, England	8.2	3.0[f]	+5.2	213	Bog	S_c, S_s, S_{sh}
Tropical: angiosperm mostly evergreen forest							
	Rio Negro, Brazil	5.6	4.7[g]	+0.9	200	Trp	S_{sh}, M_{sh}, M_g
Tundra: low shrub vegetation							
	Kuokkel area, Sweden (23)	1.15[m]	0.97[m]	+0.18	75	Ca, Em, B	M_{sh}

			Phosphorus				
Temperate: mostly angiosperm and deciduous forest							
	Coshocton, OH, U.S.	0.18[n]	0.05	+0.13	89	L, Pin	S_{sh}, S_c, S_s
	Hubbard Brook, U.S.	0.036	0.019	+0.017	130	Ac, F, B	I_g, M_{sh}
	Pago Catchment, Australia	0.33	0.26	+0.07	150	E	M_{sh}, I_g
	Silverstream, New Zealand	0.2	0.03	+0.2	135	N	S_s, S_{sh}, M_s, M_{sh}
	Taughannock Creek, NY, U.S.	0.07	0.20	−0.13	96	Ac, Ti, Ts	S_{sh}, S_c
	Walker Branch, TE, U.S.	0.54	0.02	+0.52	155	Q, C	S_c
Temperate: mostly coniferous and evergreen forest							
	Blue Range Catchment, Australia	0.39	0.42	−0.03	132	Pin	I_b, M_{sh}
	Boundary Waters Canoe Area, MIN, U.S. (24)	0.14	0.015	+0.13	70	Pin, Pic, B	I_g, S_{sh}
	Carnation Creek, Vancouver Island, Canada	0.11	0.05	+0.06	315	Ab, Ts, Th	M_b, M_s, M_c
	Cedar River, WA, U.S.	[o]	0.02	+?	136	Ps	S_s
	Clear Lake, Ontario, Canada (25)	0.35	0.09	+0.26	90	Ac, F, Q	I_g, M_g
	ELA, Ontario, Canada	0.32	0.05	+0.27	83	Pin, Pic	I_g, M_g
	Finland	0.1	0.3	−0.2	57	Pin, Pic, B	I_g, M_g
	Storsjön, Sweden	0.14	0.02	+0.12	87	Pin, Pic	I_g
	Western Cascades Range, OR, U.S.	0.29[p]	0.51[p]	−0.22	200	Ps	I_b
Temperate: bog vegetation							
	Rough Sike Catchment, England	0.6	0.4[f]	+0.2	213	Bog	S_c, S_s, S_{sh}
Tropical: angiosperm, mostly evergreen forest							
	Rio Negro, Brazil	0.2	0.1[g]	+0.1	213	Trp	S_{sh}, M_{sh}, M_g
Tundra: low shrub vegetation							
	Kuokkel area, Sweden	0.055[m]	0.041[m]	+0.014	75	Ca, Em, B	M_{sh}

(*continued*)

TABLE 24. Nutrient Budgets for Various Terrestrial Ecosystems of the World (kg/ha·yr) (*continued*)

	Location[a]	Precipitation input	Stream-water output	Net gain or loss	Annual precipitation, cm	Dominant vegetation[b]	Geology[c]
			Sulfate–sulfur				
Temperate: mostly angiosperm and deciduous forest							
	Hubbard Brook, U.S.	18.8[j]	17.6	+1.2	130	Ac, F, B	I_g, M_{sh}
	S.E. U.S.	8	7	+1	127	Q, Pin	S_s, S_{sh}, M_{sh}
	Silverstream, New Zealand	7	13	−6	135	N	S_s, S_{sh}, M_s, M_{sh}
	Taughannock Creek, NY, U.S.	21	38	−17	96	Ac, Ti, Ts	S_{sh}, S_c
	Walker Branch, TE, U.S.	18.8	11.3	+7.5	155	Q, C	S_c
Temperate: mostly coniferous and evergreen forest							
	Birkenes Watershed, Norway	15.6	26.9	−11.3	134	B, Pin, Pic	I_g
	Carnation Creek, Vancouver Island, Canada	8.7	28	−19	315	Ab, Ts, Th	M_b, M_s, M_c
	ELA, Ontario, Canada	3.0	3.2	−0.2	83	Pin, Pic	I_g, M_g
	Finland	1.4	4.7	−3.3	57	Pin, Pic, B	I_g, M_g
	Storsjön, Sweden	11.5	25.3	−13.8	87	Pin, Pic	I_g
	Velen, Sweden	10.3[e]	9.4[e]	+0.9	72	Pin, Pic	M_g, I_g

[a](1) National Academy of Sciences (1974). (2) Present study. (3) Woodwell and Whittaker (1967). (4) P. M. Hallam, personal communication (1976). (5) Gambell and Fisher (1966). (6) Miller (1963). (7) Likens (1972). (8) Gosz (1977). (9) G. Henderson, personal communication (1976). (10) Gjessing et al. (1976). (11) Scrivner (1975). (12) Cole et al. (1967). (13) Schindler et al. (1976). (14) Viro (1953). (15) W. Dickson, personal communication (1976). (16) Eriksson (1974). (17) Crisp (1966). (18) Jordan and Kline (1972). (19) Ungemach (1967, 1970). (20) Cryer (1976). (21) Taylor et al. (1971). (22) Fredriksen (1975). (23) M. Jansson, personal communication (1976). (24) Wright (1974). (25) Schindler and Nighswander (1970).

[b]Vegetation footnotes: Ab, *Abies;* Ac, *Acer;* B, *Betula;* Bog, bog species; C, *Carya; Ca, Calluna;* E, *Eucalptus;* Em, *Empetrum;* F, *Fagus;* J, *Juniperus;* L, locust; N, *Nothofagus;* Pin, *Pinus;* Pic, *Picea;* Ps, *Pseudotsuga;* Po, *Populus;* Q, *Quercus;* Th, *Thuja;* Ti, *Tilia;* Ts, *Tsuga;* Trp, tropical species.

[c]Geology footnotes: I_g, igneous, granitic; I_b, igneous, basaltic; S_c, sedimentary, carbonate; S_{sh}, sedimentary, shale; S_s, sedimentary, sandstone; M_x, metamorphic (x refers to subscripts used above).

[d]To water table.

[e]Average for 6 yr, 1968–1973.

[f]Transport of eroded peat not included.

[g]Assuming measurements of discharge and concentration are representative of the entire year.

[h]Calculated from Mg inputs and Na:Mg ratio in seawater.

[i]NO_3-N + NH_4-N.

[j]Also includes net gaseous uptake and aerosol deposition; see Table 10 for bulk precipitation values.

[k]NO_3-N only.

[l]Average of five drainage basins, 1971–1973.

[m]Average for 5 yr, 1971–1975.

[n]Calculated from 89 cm of precipitation times concentration (0.02 mg P/l; Taylor et al., 1971).

[o]"Trace."

[p]Average of five drainage basins, 1972–1973.

[q]Data for calcium omitted because of apparent contamination of precipitation samples.

biomes have conservative losses of nutrients relative to inputs or amounts cycled internally, and (9) Stream-water losses may be very significantly changed by disturbance of the ecosystem, such as forest cutting or other land management practices. The array of biogeochemical responses to clear cutting and recovery of the northern hardwood ecosystem at Hubbard Brook is discussed in our book *Patterns and Processes in a Forested Ecosystem* (Bormann and Likens, 1979; see also Likens and Bormann, 1974a).

8 Summary Discussion and Conclusions

This book summarizes our current understanding of the biogeochemistry of a northern hardwood forest ecosystem at Hubbard Brook. It emphasizes the usefulness of the small watershed technique of biogeochemical measurement in shedding light on ecosystem function.

Trees are long lived and over time accumulate large amounts of biomass as wood; it is therefore not surprising that trees dominate the forest landscape and are used to name the ecosystem types (plant associations or biomes). Humans have long looked to the forests for firewood and timber, as well as for clean water, game, and recreation, but their demands often have been at variance with the structure of the forest ecosystem and large areas have been (and are being) cleared of forest vegetation. This is currently proceeding at an unprecedented rate in the tropical forest ecosystems (e.g., Croat, 1972; Gómez-Pompa et al., 1972).

Water in some form is also a conspicuous feature of the humid forested landscape. Rain and snow not only supply the water that replenishes ground water reserves and fills stream and river channels, ponds, lakes, swamps, and marshes but also provide the moisture and many of the nutrients for forest growth. Chemicals, particularly those without a prominent gaseous phase, such as calcium, magnesium, or potassium, are transported largely into and out of ecosystems by moving water.

In forested landscapes much of the water that falls to the land's surface as precipitation is first intercepted by the leaves and branches of the trees and herbaceous layer and then by litter on the ground. Therefore, the kinetic energy and chemical composition of the water may be appreciably altered before the water comes into contact with the mineral soil. This both minimizes erosion and regulates nutrient flow. This ostensibly simple and obvious interaction emphasizes that the regulation of ecosystem processes, such as biogeochemical flux, depends on the integrity of the whole ecosystem.

Water that has fallen to the surface may run off overland (minimal in most forests), infiltrate into the soil, or evaporate. Again, the biotic portion of the forest ecosystem may alter both the distribution of potential and kinetic energy and the chemistry of water circulating within an ecosystem.

There are several ways, then, in which water may affect the biogeochemistry of a forested ecosystem. We have found it useful for quantitative studies to select ecosystems that (1) are watersheds, (2) have a watertight bedrock or other substrate, and (3) have a uniform biogeochemical environment (Bormann and Likens, 1967; Likens and Bormann, 1972).

The flux of water and nutrients across an ecosystem's boundaries, as well as internal cycling, is vital for the maintenance of a natural ecosystem. Environmental stress, such as pollution, deforestation, or road building, will alter these conditions of nutrient flux and cycling and thereby disrupt the function of the ecosystem. Too little is known about the long-term effects of such alterations or about the ability of an ecosystem to respond, recover, or improvise following disturbance.

The yield of liquid water in streamflow is an important consideration to watershed managers. For example, Douglass and Swank (1972) state that "Because the difference between precipitation input and vapor loss represents the quantity of water available for man's use, the watershed manager seeks to reduce the total vapor loss from the forest vegetation in order to increase the flow of streams" (p. 1). However, because liquid water that flows out of an ecosystem contains nutrients (whether at a constant concentration or not) it may be argued that transpiration is a nutrient conservation mechanism for the ecosystem. Moreover, as precipitation comes in contact with the vegetation its chemistry is altered significantly. The kinetic energy of the falling raindrop is largely absorbed by the vegetation canopy during the growing season and the potential for erosion is reduced. Litter and roots in the forest soil also reduce erosion. The change of liquid water to vapor by the ecosystem also may have great impact in reducing the potential for erosion from liquid water. Therefore, the biotic structure of these forest ecosystems may alter both the potential energy and the chemistry of water as it passes through the ecosystem. Inadequate consideration has been given to the interplay of such changes in the hydrologic cycle on nutrient flux for forest ecosystems. This is of particular concern when the basic structure of the forest ecosystem is altered, e.g., by vegetation removal (Likens et al., 1970).

There is currently considerable effort to quantify biogeochemical relationships for various kinds of terrestrial ecosystems. As a result of our study, we emphasize that:

(1) Considerable thought be given (i) to selection of the ecosystem, including boundaries, for study–millions of dollars and thousands of hours of time may be wasted by a poor initial choice; and (ii) to methodology, such as frequency and location of sampling;

criteria for acceptance or rejection of particular samples; and types and reliability of chemical analyses. This is important because small analytical mistakes can easily be blown out of proportion in the multiplication procedures that are inevitable in the construction of ecosystem budgets.

(2) Ecosystems are open. Water and nutrients continually flux through the boundaries and cycle internally between the various components of the ecosystem. Studies at Hubbard Brook have demonstrated the importance of the abiotic factors and quantitative mass balance studies to understanding the structure and function of ecosystems.

(3) Long-term records are important in identifying patterns of nutrient flux through ecosystems. Flux is greatly influenced by hydrology and, during 11 yr precipitation has ranged from 95 to 186 cm/yr at the HBEF. Measurements in either of these extreme years would have given poor estimates of average conditions. Over a 10-yr period, moreover, annual nitrate and hydrogen ion inputs in precipitation increased 2.3 and 1.4 times, respectively. The output of NO_3 in stream water also increased during this period and seemed to be strongly influenced by soil freezing during two of the 11 winters.

(4) Although the chemistry of precipitation is highly variable and dependent on complex meterorologic conditions, meteorologic input is an important source of chemicals for the ecosystem. For S, N, Mg, Na, Ca, and K, meteorologic input supplies about 66, 26, 6, 5, 4, and 1%, respectively, of the annual uptake by green plant biomass.

(5) Meteorologic input may be strongly influenced by advertent and inadvertent manipulation by humans. Acid precipitation is a good example: Most acidity in the rain and snow at Hubbard Brook is traceable to sulfur pollution but during 1964–1974 increases in the input of acid precipitation seem to be primarily related to increases in nitrate resulting from internal combustion engines.

(6) Currently the ecosystem is acting as a filter for atmospheric contaminants, such as H^+, N, S, P, and certain heavy metals, which accumulate within the ecosystem. The forest ecosystem, therefore, is currenty acting as a pollution buffer for society; the ultimate cost of this activity is unknown but it seems to trend toward long-term ecosystem degradation.

(7) Annual evapotranspiration (mostly transpiration from vegetation) at Hubbard Brook is essentially constant over a wide range of precipitation and environmental conditions. Water converted to vapor by evapotranspiration cannot serve to erode the system or transport dissolved substances out of the ecosystem. Evapotranspiration not only serves as a regulator for the hydrologic

cycle, therefore, but also affects nutrient flux as well. Moreover, the forest ecosystem exercises a strong regulatory function over the chemistry of liquid water passing through it. This is achieved because the structure of the ecosystem causes all precipitation to come into intimate contact with, first, the living and dead biomass and then the minerals and organic matter in the soil profile. Rapid exchange reactions in each of these compartments govern the chemical composition of liquid water leaving the ecosystem.

(8) Evapotranspiration reduces the volume of liquid water in the ecosystem. Streamwater concentrations could be increased by 1.6 times in this way. If evapotranspiration were the only operational factor, the output (concentration $\times$ amount of water) would equal the input (1.33×10^3 Eq/ha·yr). Instead, the ecosystem is much more complex. Nutrients are released internally by weathering, accumulated by vegetation, etc. The result at Hubbard Brook is that the ionic strength of stream water (0.20 mEq/l) is twice that of precipitation (0.10 mEq/l) and the average output is 1.65×10^3 Eq/ha·yr, whereas the average input is 1.35×10^3 Eq/ha·yr. In addition, the proportions of dissolved species change from precipitation to streamflow, further illustrating the importance of biologic and chemical reactions occurring within the ecosystem.

Hydrogen ion and sulfate dominate precipitation (pH $\sim$ 4), whereas calcium and sulfate dominate stream-water (pH $\sim$ 5) chemistry. Nitrate input in precipitation is only twice as large on an equivalent basis as ammonium but in stream-water output the proportion increases to 14:1.

(9) Stream-water chemistry of the forest ecosystem is highly predictable. Concentrations of Na^+ and SiO_2 are predictably diluted as streamflow rates increase, whereas Al^{3+}, NO_3^-, H^+, K^+, and dissolved organic carbon are increased as streamflow increases. However, changes in concentrations are relatively small in relation to the great range in flow rates. It is well known that some fixed time series for sampling may seriously underestimate or overestimate a highly variable parameter (e.g., Claridge, 1970); however, most of the dissolved chemicals in stream water at the HBEF have such constant concentrations that serious errors are not produced by weekly or even monthly sampling in some cases (Johnson et al., 1969; Likens et al., 1967). Concentrations of both nitrate and potassium in stream water are highly dependent on seasonal biologic activity within the ecosystem. The biogeochemical reaction rates within these ecosystems are rapid; therefore, stream-water chemistry quickly and clearly reflects environmental conditions within the forest ecosystem.

(10) As a consequence of concentration-flow rate relations of streams, the output of most individual nutrients and the total load of dissolved substances can be closely predicted from the annual output of water alone. The few calculated values in Table 11 are therefore very reliable estimates, because hydrology has been measured and only the chemical concentration has been estimated.

(11) The aggrading northern hardwood ecosystem is very resistant to particulate matter erosion. This is because of its geology (stoney till) and highly efficient mechanisms associated with living and dead biomass. Particulate matter erosion is highly dependent on flow rate, in contrast to dissolved substance removal, which is independent of flow rate. Occasional, individual storms with high discharges do most of the erosion and transport of particulate matter. In contrast to all of the other nutrients, the bulk of iron and phosphorus is lost from watershed ecosystems at Hubbard Brook in the particulate form. The stream-water output for these two elements therefore may reflect more closely the rate of stream discharge than the amount of annual streamflow. Samples of stream water obtained according to a standard time series therefore may be less accurate in estimating stream-water concentrations of iron and phosphorus than for substances occurring primarily in the dissolved form. Because discharge affects the concentration of dissolved substances and particulate matter differently, any management of water quality or total export of these materials requires different approaches at different levels of discharge.

(12) In particulate matter removal, bedload accounts for slightly more than 50% of the material removed. Many studies only measure suspended matter and hence tend to underestimate the total output of particulate matter.

(13) Dissolved substances losses from the ecosystem (110 kg of ionic substances, 38 kg of dissolved silica, and 17 kg of dissolved organic matter per hectare·year) are about five times greater than particulate matter losses (11 kg of organic matter plus 22 kg of inorganic particles). Total losses of dissolved substances and particulate matter (198 kg/ha·yr) in stream water are about 1.5 times greater than total inputs in bulk precipitation (134 kg/ha·yr).

(14) The lowering of the Hubbard Brook landscape, which is dominated by forests, is primarily caused by solution losses and, to a lesser extent, by particulate matter losses. Slow mass movements, such as soil creep, may deliver materials to stream beds, where they are removed. At higher elevations and with steeper slopes, debris avalanches also may play a major role in lowering

the landscape. Another factor that may be of considerable importance is disturbances that temporarily destroy the forest and make it more vulnerable to both solution and particulate matter losses.

(15) In establishing input-output budgets for individual nutrients both direction of change, i.e., whether there is a net loss or gain, and magnitude of change are important. Quantitative data from six adjacent watersheds provided estimates of variability for annual budgets with time. For some ions, these budget parameters may be established with one or a few years of data, for example, direction: $Ca^{2+}(-)$, $Mg^{2+}(-)$, $Na^{+}(-)$, $Al^{3+}(-)$, $NH_4^{+}(+)$, $H^{+}(+)$, $SO_4^{2-}(-)$, $PO_4^{3-}(+)$, dissolved silica $(-)$, $HCO_3^{-}(-)$. For other ions, K^{+}, NO_3^{-} and Cl^{-}, one or a few years gave unreliable results for direction and magnitude. For some budgets, therefore, longer term studies are necessary to obtain valid budgetary data.

(16) Aggrading ecosystems at Hubbard Brook are accumulating significant amounts of nitrogen by microbial nitrogen fixation or other inputs of nitrogenous gases or aerosols according to estimates made by the small watershed technique. Field studies, using the acetylene reduction method, indicate vigorous nitrogen fixation in dead wood.

(17) A sulfur balance established using the Hubbard Brook ecosystem model and data gathered by the small watershed technique indicate that substantial amounts of dry deposition (6.1 kg/ha·yr) are entering the Hubbard Brook ecosystem, even though these watersheds are located more than 100 km from major polluting industries or urban areas.

(18) Three input-output patterns emerge when small watershed data are examined on a monthly basis: output exceeds input in every month for Mg^{2+}, Ca^{2+}, Na^{+} and Al^{3+}; output is less than input for NH_4^{+}, H^{+}, and PO_4^{3-}; and crossover patterns, where for a time output exceeds input and then output is less than input, occur for K^{+}, SO_4^{2-}, Cl^{-}, and NO_3^{-}. These patterns are distributed among elements with both sedimentary and atmospheric cycles and reflect a complex interaction involving such factors as precipitation input, biologic activity, and climatic variations.

(19) Not only must net losses of ions from the ecosystem be included in a calculation of weathering rates, but those built into any net accumulation of biomass must also be included. At Hubbard Brook failure to include annual biomass accretion of nutrients results in a weathering estimate that is 50% too low.

(20) Chemical weathering of minerals within the ecosystem is driven by protons derived from hydrogen ions in precipitation and those generated internally within the system by biogeochemical re-

actions involving carbon, nitrogen, and sulfur. The cationic denudation rate is about 2.0×10^3 Eq/ha·yr. We calculate that about 50% of the H^+ ions derived during the weathering reaction is derived from external sources, i.e., acid precipitation, and 50% is from internal biogeochemical sources. However, this estimate is based on the assumption that all H^+ ions, external plus internal, are consumed in weathering reactions.

(21) When compared to the percentage occurrence of elements in the weathering substrate, weathering estimates, indicate that differential weathering is occurring. Most rapid to least rapid are Ca > Na > Mg > K > Al > Si. The ecosystem is becoming relatively enriched in reverse order.

(22) Complete nutrient budgets for the Hubbard Brook ecosystem reveal certain basic features of nutrient cycling not otherwise apparent. These are: (a) Wet and dry deposition are major sources of nutrients, particularly for sulfur, nitrogen, chloride, and phosphorus. (b) Weathering is a major source for Ca, Mg, K, and Na. (c) Biologic activity, photosynthesis, and nitrogen fixation, play a major role in the input of carbon and nitrogen and are important (impaction surfaces plus gaseous uptake) for sulfur. (d) The aggrading northern hardwood forest shows absolute gains for N, S, P, and Cl and losses for Si, Ca, Na, Al, Mg, and K. The latter losses are made up by weathering of primary minerals within the ecosystem. Weathering substrate may be decreasing with time.

(23) Calcium, an example of a sedimentary cycle, tends to be held within the system, with net losses being but a small proportion of the calcium on the exchange sites, 2%. However, net losses are 19% as large as annual uptake by green plants. Although sulfur has a gaseous cycle, some 99% of the sulfur in the ecosystem is found in the soil. The ecosystem approach provides an estimate of net gaseous uptake and aerosol deposition of sulfur. These sources account for about 32% of the input for the system. Meteorologic inputs dominate, in general, and release by weathering is small. Net losses of sulfur in streamflow are only 4% of the annual plant uptake.

(24) Many of the important biogeochemical relationships within the forest ecosystem are regulated by microorganisms. Unfortunately, we know relatively little about the transformations mediated by microbes, such as nitrogen fixation, nitrification, and denitrification or sulfur oxidation and reduction, at Hubbard Brook. These aspects of nutrient cycling as they relate to nutrient flux must be more carefully evaluated in future studies.

(25) Input-output budgets for a diverse series of relatively undisturbed, vegetated watersheds throughout the world have many

patterns in common with Hubbard Brook. Although the magnitude of precipitation, geologic substrate, vegetation type, and proximity to marine or anthropogenic emissions may vary greatly for these widespread and diverse watersheds, the constant pattern that emerges from these data is remarkable. Inputs of phosphorus and nitrogen in bulk precipitation are generally greater than losses in drainage waters on an annual basis. Conversely, outputs of calcium, magnesium, sodium, and potassium are generally greater than precipitation inputs. Net losses of potassium are small compared to inputs. The magnitude and direction for sulfate budgets are variable and depend largely on geologic substrate and intensity of air pollution.

(26) Ecosystem studies are now at the stage to provide quantitative answers for important ecological questions. The small watershed technique is useful in providing an experimental framework for these studies, which must consider geologic and meteorologic aspects as well as biologic features. It can provide a similar framework in experimentally designed mesocosm studies (e.g., The Sandbox Experiment, Bormann et al., 1987, 1993). Such questions include: (a) How do ecosystems change with time? Do they become more or less efficient in cycling and storing nutrients? (b) What are the effects of disturbance on the biogeochemistry of ecosystems? (c) How would weathering rates change in relation to changes in rates of biologic activity? Are there conditions under which weathering does not occur at all? (d) Of critical importance is the finding that some of the ions produced by weathering do not move downhill but are instead accumulated in the growing vegetation. If the forest were to reach some kind of steady state of growth, what would happen to the biogeochemistry of a complex ecosystem?

Epilog

Aldo Leopold was a very wise and perceptive observer. Some might have called him a conservationist; others might have called him an ecologist or a biogeochemist. Actually he was a brilliant, yet humble student of nature in the purest sense. Much of what this little book has described in quantitative terms Aldo Leopold outlined in general terms 45 yr ago. His insight and prose have been an inspiration to us for a long time. We therefore think it appropriate for Aldo Leopold to have the last word in our book and so, from *A Sand County Almanac* (pp. 104–108) (1949), the following:

Odyssey

X had marked time in the limestone ledge since the Paleozoic seas covered the land. Time, to an atom locked in a rock, does not pass.

The break came when a bur-oak root nosed down a crack and began prying and sucking. In the flush of a century the rock decayed, and X was pulled out and up into the world of living things. He helped build a flower, which became an acorn, which fattened a deer, which fed an Indian, all in a single year.

From his berth in the Indian's bones, X joined again in chase and flight, feast and famine, hope and fear. He felt these things as changes in the little chemical pushes and pulls that tug timelessly at every atom. When the Indian took his leave of the prairie, X moldered briefly underground, only to embark on a second trip through the bloodstream of the land.

This time it was a rootlet of bluestem that sucked him up and lodged him in a leaf that rode the green billows of the prairie June, sharing the common task of hoarding sunlight. To this leaf also fell an uncommon task: flicking shadows across a plover's eggs. The ecstatic plover, hovering overhead, poured praises on something perfect: perhaps the eggs, perhaps the shadows, or perhaps the haze of pink phlox that lay on the prairie.

When the departing plovers set wing for the Argentine, all the bluestems waved farewell with tall new tassels. When the first geese came out of the north and all the bluestems glowed wine-red, a forehanded deermouse cut the leaf in which X lay, and buried it in an underground nest, as if to hide a bit of Indian summer from the thieving frosts. But a fox detained the mouse, molds and fungi took the nest apart, and X lay in the soil again, foot-loose and fancy-free.

Next he entered a tuft of side-oats grama, a buffalo, a buffalo chip, and again the soil. Next a spiderwort, a rabbit, and an owl. Thence a tuft of sporobolus.

All routines come to an end. This one ended with a prairie fire, which reduced the prairie plants to smoke, gas, and ashes. Phosphorus and potash atoms stayed in the ash, but the nitrogen atoms were gone with the wind. A spectator might, at this point, have predicted an early end of the biotic drama, for with fires exhausting the nitrogen, the soil might well have lost its plants and blown away.

But the prairie had two strings to its bow. Fires thinned its grasses, but they thickened its stand of leguminous herbs; prairie clover, bush clover, wild bean, vetch, lead-plant, trefoil, and Baptisia, each carrying its own bacteria housed in nodules on its rootlets. Each nodule pumped nitrogen out of the air into the plant, and then ultimately into the soil. Thus the prairie savings bank took in more nitrogen from its legumes than it paid out to its fires. That the prairie is rich is known to the humblest deermouse; why the prairie is rich is a question seldom asked in all the still lapse of ages.

Between each of his excursions through the biota, X lay in the soil and was carried by the rains, inch by inch, downhill. Living plants retarded the wash by impounding atoms; dead plants by locking them to their decayed tissues. Animals ate the plants and carried them briefly uphill or downhill, depending on whether they died or defecated higher or lower than they fed. No animal was aware that the altitude of his death was more important than his manner of dying. Thus a fox caught a gopher in a meadow, carrying X uphill to his bed on the brow of a ledge, where an eagle laid him low. The dying fox sensed the end of his chapter in foxdom, but not the new beginning in the odyssey of an atom.

An Indian eventually inherited the eagle's plumes, and with them propitiated the Fates, whom he assumed had a special interest in Indians. It did not occur to him that they might be busy casting dice against gravity; that mice and men, soils and songs, might be merely ways to retard the march of atoms to the sea.

One year, while X lay in a cottonwood by the river, he was eaten by a beaver, an animal that always feeds higher than he dies. The beaver starved when his pond dried up during a bitter frost. X rode the carcass down the spring freshet, losing more altitude each hour than heretofore in a century. He ended up in the silt of a backwater bayou, where he fed a crayfish, a coon, and then an Indian, who laid him down to his last sleep in a mound on the riverbank. One spring an oxbow caved the bank, and after one short week of freshet X lay again in his ancient prison, the sea.

An atom at large in the biota is too free to know freedom; an atom back in the sea has forgotten it. For every atom lost to the sea, the prairie pulls another out of the decaying rocks. The only certain truth is that its creatures must suck hard, live fast, and die often, lest its losses exceed its gains.

* * *

It is the nature of roots to nose into cracks. When Y was thus released from the parent ledge, a new animal had arrived and begun redding up the prairie to fit his own notions of law and order. An oxteam turned the prairie sod, and Y began a succession of dizzy annual trips through a new grass called wheat.

The old prairie lived by the diversity of its plants and animals, all of which were useful because the sum total of their co-operations and competitions achieved continuity. But the wheat farmer was a builder of categories; to him only wheat and oxen were useful. He saw the useless pigeons settle in clouds upon his wheat, and shortly cleared the skies of them. He saw the chinch bugs take over the stealing job, and fumed because here was a useless thing too small to kill. He failed to see the downward wash of over-wheated loam, laid bare in spring against the pelting rains. When soil-wash and chinch bugs finally put an end to wheat farming, Y and his like had already traveled far down the watershed.

When the empire of wheat collapsed, the settler took a leaf from the old prairie book: he impounded his fertility in livestock, he augmented it with nitrogen-pumping alfalfa, and he tapped the lower layers of the loam with deep-rooted corn.

But he used his alfalfa, and every other new weapon against wash, not only to hold his old plowings, but also to exploit new ones which, in turn, needed holding.

So, despite alfalfa, the black loam grew gradually thinner. Erosion engineers built dams and terraces to hold it. Army engineers built levees and wing-dams to flush it from the rivers. The rivers would not flush, but raised their beds instead, thus choking navigation. So the engineers built pools like gigantic beaver ponds, and Y landed in one of these, his trip from rock to river completed in one short century.

On first reaching the pool, Y made several trips through water plants, fish, and waterfowl. But engineers build sewers as well as dams, and down them comes the loot of all the far hills and the sea. The atoms that once grew pasque-flowers to greet the returning plovers now lie inert, confused, imprisoned in oily sludge.

Roots still nose among the rocks. Rains still pelt the fields. Deermice still hide their souvenirs of Indian summer. Old men who helped destroy the pigeons still recount the glory of the fluttering hosts. Black and white buffalo pass in and out of red barns, offering free rides to itinerant atoms.

References

Anderson, D.H., and H.E. Hawkes. 1958. Relative mobility of the common elements in weathering of some schist and granite areas. *Geochim. Cosmochim. Acta*, **14**(3):204–210.

Arefýeva, Z.N., and B.D. Kolesnikof. 1964. Chemistry and biochemistry dynamics of ammonia and nitrate nitrogen in forest soils of the Transurals at high and low temperatures. *Sov. Soil Sci.*, **3**:246–260.

Art, H.W. 1976. Ecological Studies of the Sunken Forest Island National Seashore, New York. National Park Service. Scientific Monograph Series, No. 7.

Art, H.W., F.H. Bormann, G.K. Voigt, and G.M. Woodwell. 1974. Barrier Island forest ecosystem: the role of meteorological nutrient inputs. *Science*, **184**:60–62.

Bailey, S.W. 1994. Biogeochemistry of aluminum and calcium in a linked forest-aquatic ecosystem. Ph.D. Thesis, Syracuse University, Syracuse, NY. 86 pp.

Barrett, E., and G. Brodin. 1955. The acidity of Scandinavian precipitation. *Tellus*, **7**:251–257.

Barton, C.C., and S.W. Bailey. 1995. Bedrock Geologic Map of the Hubbard Brook Valley, New Hampshire. (In preparation).

Beamish, R.J. 1976. Effects of precipitation on Canadian lakes. pp. 479–498 In: *Proc. of the First International Symposium on Acid Precipitation and the Forest Ecosystem*. L.S. Dochinger and T.A. Seliga (eds.). U.S.D.A. For. Serv. Gen. Tech. Rep. NE-23.

Billings, M.P. 1956. *The Geology of New Hampshire*. Part II. *Bedrock Geology*. New Hampshire State Planning and Development Commission, Concord, NH. 200 pp.

Bolin, B. (ed.). 1971. *Report of the Swedish Preparatory Committee for the United Nations Conference on Human Environment*. Norstedt and Söner, Stockholm, Sweden.

Bormann, B.T., F.H. Bormann, W.B. Bowden, R.S. Pierce, S.P. Hamburg, D. Wang, M. Snyder, C.Y. Li, and R.C. Ingersoll. 1993. Rapid N_2 fixation in pines, alder, and locust: evidence from the sandbox ecosystem study. *Ecology*, **74**(2):583–598.

Bormann, F.H. 1974. Acid rain and the environmental future. *Environ. Cons.*, p. 270.

Bormann, F.H., and G.E. Likens. 1967. Nutrient cycling. *Science*, **155**:424–429.

Bormann, F.H., and G.E. Likens. 1979. *Pattern and Process of a Forested Ecosystem*. Springer-Verlag, New York, NY. 253 pp.
Bormann, F.H., G.E. Likens, and J.S. Eaton. 1969. Biotic regulation of particulate and solution losses from a forested ecosystem. *BioScience*, **19**(7):600–610.
Bormann, F.H., T.G. Siccama, G.E. Likens, and R.H. Whittaker. 1970. The Hubbard Brook Ecosystem Study: composition and dynamics of the tree stratum. *Ecol. Monogr.*, **40**(4):373–388.
Bormann, F.H., G.E. Likens, T.G. Siccama, R.S. Pierce, and J.S. Eaton. 1974. The effect of deforestation on ecosystem export and the steady-state condition at Hubbart Brook. *Ecol. Monogr.*, **44**(3):255–277.
Bormann, F.H., G.E. Likens, and J. Melillo. 1977. Nitrogen budget for an aggrading northern hardwood forest ecosystem. *Science*, **196**(4293):981–983.
Bormann, F.H., W.D. Bowden, R.S. Pierce, S. Hamburg, G.K. Voigt, R.C. Ingersoll, and G.E. Likens. 1987. The Hubbard Brook sandbox experiment. pp. 251–256. In: *Restoration Ecology*. W.R. Jordan, M.E. Gilpin and J.D. Aber (eds.) Cambridge University Press, Cambridge, England.
Bradley, E., and R.V. Cushman. 1956. Memorandum report on geologic and ground-water conditions in the Hubbard Brook watershed, New Hampshire. On file at Northeastern Forest Experiment Station. Durham, NH. 15 pp.
Brady, N.C. 1974. *The Nature and Properties of Soils*. Macmillan, New York. 639 pp.
Braun, E.L. 1950. *Deciduous Forests of Eastern North America*. The Blakiston Co., Philadelphia, PA. 594 pp.
Burton, T.M., and G.E. Likens. 1975. Energy flow and nutrient cycling in salamander populations in the Hubbard Brook Experimental Forest, New Hampshire. *Ecology*, **56**(5):1068–1080.
Butler, T.J., and G.E. Likens. 1991. The impact of changing regional emissions on precipitation chemistry in the eastern United States. *Atmos. Environ.*, **25A**:305–315.
Carlisle, A., A.H.F. Brown, and E.J. White. 1967. The nutrient content of tree stem flow and ground flora litter and leachates in a sessile oak (*Quercus petraea*) woodland. *J. Ecol.*, **55**:615–627.
Carroll, D. 1970. *Rock Weathering*. Plenum Press, New York, NY. 203 pp.
Claridge, G.G.C. 1970. Studies in element balances in a small catchment at Taita, New Zealand. *Proceedings of the International Association of Scientific Hydrology, UNESCO Symp.*, on Results of Research on Representative and Experimental Basins. pp. 523–540. Wellington, New Zealand, December 1970.
Cogbill, C.V. 1976. The effect of acid precipitation on tree growth in eastern North America. pp. 1027–1032. In: *Proceedings of the First International Symposium on Acid Precipitation and the Forest Ecosystem*. L.S. Dochinger, and T.A. Seliga (eds.) U.S.D.A For. Serv. Gen. Tech. Rep. NE-23.
Cogbill, C.V., and G.E. Likens. 1974. Acid precipitation in the northeastern United States. *Water Resources Res.*, **10**:1133–1137.
Cole, D.W., S.P. Gessel, and S.F. Dice. 1967. Distribution and cycling of nitrogen, phosphorus, potassium, and calcium in a second-growth douglas-fir ecosystem. *Symposium on Primary Productivity and Mineral Cycling in Natural Ecosystems*. New York City, December 27, 1967, University of Maine Press, Orono, ME. 245 pp.

Conrad, V. 1941. The variability of precipitation. *Mon. Weath. Rev. Wash.*, **69**:5.
Covington, W.W. 1975. Successional dynamics of organic matter and nutrient content of forest floor and leaf fall in northern hardwoods. *Bull. Ecol. Soc. Amer.*, **56**(2):12.
Covington, W.W. 1976. Secondary succession in northern hardwoods: forest floor organic matter and nutrients and leaf fall. Ph.D. Thesis, Yale University, New Haven, CT. 117 pp.
Cowling, D.W., L.H.P. Jones, and D.R. Lockyer. 1973. Increased yield through correction of sulphur deficiency in ryegrass exposed to sulphur dioxide. *Nature*, **243**(5407):479–480.
Crisp, D.T. 1966. Input and output of minerals for an area of pennine moorland: the importance of precipitation, drainage, peat erosion and animals. *J. Appl. Ecol.*, **3**:327–348.
Croat, T.B. 1972. The role of overpopulation and agricultural methods in the destruction of tropical ecosystems. *BioScience*, **22**(8):465–467.
Cryer, R. 1976. The significance and variation of atmospheric nutrient inputs in a small catchment system. *J. Hydrol.*, **29**:121–137.
Dominski, A.S. 1971. Nitrogen transformations in a northern-hardwood podzol on cutover and forested sites. Ph.D. Thesis, Yale University, New Haven, CT. 157 pp.
Douglass, J.E., and W.T. Swank. 1972. Streamflow modification through management of Eastern forests. Southeast Forest Experiment Station, U.S.D.A. Forest Service Research Paper SE-94, Washington, DC. 15 pp.
Driscoll, C.T., N. van Breemen, and J. Mulder. 1985. Aluminum chemistry in a forested spodosol. *Soil Sci. Soc. Amer. J.*, **49**(2):437–444.
Driscoll, C.T., G.E. Likens, L.O. Hedin, J.S. Eaton, and F.H. Bormann. 1989. Changes in the chemistry of surface waters: 25-year results at the Hubbard Brook Experimental Forest, NH. *Environ. Sci. Technol.*, **23**(2):137–143.
Duvigneaud, P., and S. Denaeyer-DeSmet. 1964. Le cycle des éléments biogènes dans l'écosystème forêt (Forêts tempèrées caducifoliées). *Lejeunia*, **28**: 1–148.
Eaton, J.S., G.E. Likens, and F.H. Bormann. 1973. Throughfall and stemflow chemistry in a northern hardwood forest. *J. Ecol.*, **61**:495–508.
Eaton, J.S., G.E. Likens, and F.H. Bormann. 1976. The biogeochemistry of sulfur in a northeastern forest ecosystem. p. 43. In: *The Tenth Middle Atlantic Regional Meeting Abstracts*, American Chemical Society, February 23–26, Philadelphia, PA.
Eriksson, E. 1952. Composition of atmospheric precipitation. II. Sulfur, chloride, iodine compounds. Bibliography. *Tellus*, **4**:280–303.
Eriksson, E. 1974. Vattnet-Kemikaliebäraren. *Forskning och Framsteg*, **5**:41–45.
Fahey, T.J., and J. Hughes. 1994. Fine root dynamics in a northern hardwood forest ecosystem, Hubbard Brook Experimental Forest, NH. *J. Ecology*, **82**: 533–548.
Federer, C.A. 1973. Annual cycles of soil and water temperatures at Hubbard Brook. U.S.D.A. Forest Service Research Note NE-167. Upper Darby. PA. 7 pp.
Federer, C.A., and D. Lash. 1978. BROOK: A hydrologic simulation model for eastern forests. Water Resour. Res. Center, Univ. of New Hampshire, Durham. Research Report No. 19, 84 pp.

Federer, C.A., L.D. Flynn, C.W. Martin, J.W. Hornbeck, and R.S. Pierce. 1990. Thirty years of hydrometeorologic data at the Hubbard Brook Experimental Forest, New Hampshire. U.S.D.A. Forest Service Gen. Tech. Rep. NE-141. 44 pp.

Fisher, D.W., A.W. Gambell, G.E. Likens, and F.H. Bormann. 1968. Atmospheric contributions to water quality of streams in the Hubbard Brook Experimental Forest, New Hampshire. *Water Resources Res.*, **4**(5):1115–1126.

Fisher, S.G. 1970. Annual energy budget of a small forest stream ecosystem: Bear Brook, West Thornton, New Hampshire. Ph.D. Thesis, Dartmouth College, Hanover, NH. 97 pp.

Fisher, S.G., and G.E. Likens. 1973. Energy flow in Bear Brook, New Hampshire: an integrative approach to stream ecosystem metabolism. *Ecol. Monogr.*, **43**(4):421–439.

Flaccus, E. 1958a. Landslides and their revegetation in the White Mountains of New Hampshire. Ph.D. dissertation, Duke University, Durham, NC. 186 pp.

Flaccus, E. 1958b. White Mountain landslides. *Appalachia*, **32**:175–191.

Fowler-Billings, K., and L.R. Page. 1942. The Geology of the Cardigan and Rumney Quadrangles, New Hampshire. New Hampshire Planning and Development Commission. Concord, NH. 31 pp.

Fredriksen, R.L. 1975. Nitrogen, phosphorus and particulate matter budgets of five coniferous forest ecosystems in the Western Cascades Range, Oregon, Ph.D. Thesis, Oregon State University, Corvallis, OR.

Galloway, J.N., G.E. Likens, and E.S. Edgerton. 1976. Acid precipitation in the northeastern United States: pH and acidity. *Science*. **194**:722–724.

Gambell, A.W., and D.W. Fisher. 1966. Chemical composition of rainfall of eastern North Carolina and southeastern Virginia. United States Geological Survey Water Supply Paper 1535-K. Washington, DC. pp. K1–K41.

Gjessing, E.T., A. Henriksen, M. Johannessen, and R.F. Wright. 1976. Effects of acid precipitation on freshwater chemistry. pp. 65–85. In: F.H. Braekke (ed.). *Impact of Acid Precipitation on Forest and Freshwater Ecosystems in Norway*. Report FR 6/76, SNSF-Project, Oslo, Norway.

Goldich, S.S. 1938. A study of rock weathering. *J. Geol.*, **46**:17–58.

Gómez-Pompa, A., C. Vázquez-Yanes, and S. Guevara. 1972. The tropical rainforest: a nonrenewable resource. *Science*, **177**:762–765.

Gorham, E. 1958. Atmospheric pollution by hydrochloric acid. *Quart. J. Roy. Meteorol. Soc.*, **84**(361):274–276.

Gosz, J.R. 1980. Nutrient budget studies of forests along an elevational gradient in New Mexico. *Ecology* **61**(3):515–521.

Gosz, J.R., G.E. Likens, and F.H. Bormann. 1972. Nutrient content of litter fall on the Hubbard Brook Experimental Forest, New Hampshire. *Ecology*, **53**(5): 769–784.

Gosz, J.R., G.E. Likens, and F.H. Bormann. 1973. Nutrient release from decomposing leaf and branch litter in the Hubbard Brook Forest, New Hampshire. *Ecol. Monogr.*, **43**(2):173–191.

Gosz, J.R., G.E. Likens, and F.H. Bormann. 1976. Organic matter and nutrient dynamics of the forest and forest floor in the Hubbard Brook Forest. *Oecologia* (Berlin), **22**:305–320.

Granat, L. 1972. On the relation between pH and the chemical composition in atmospheric precipitation. *Tellus,* **24**:550–560

Hack, J.T. and J.C. Goodlett. 1960. Geomorphology and forest ecology of a mountain region in the Central Appalachians. United States Geological Survey Professional Paper 347. Washington, DC 66 pp.

Hamon, W.R. 1961. Estimating potential evapotranspiration. *Amer. Soc. Civ. Eng. Proc.,* **87**(HY3):107–120.

Hart, G.E. Jr. 1966. Streamflow characteristics of small forested watersheds in the White Mountains of New Hampshire. Ph.D. Thesis, University of Michigan. Ann Arbor, MI. 141 pp.

Hart, G.E. Jr., R.E. Leonard, and R.S. Pierce. 1962. Leaf fall, humus depth, and soil frost in a northern hardwood forest. Research Note No. 131, Northeastern Forest Experiment Station, Upper Darby, PA. 3 pp.

Hedin, L.O., L. Granat, G.E. Likens, T.A. Buishand, J.N. Galloway, T.J. Butler, and H. Rodhe, 1994. Steep declines in atmospheric base cations in regions of Europe and North America. *Nature*, **367**:351–354.

Herman, F.A. and E. Gorham. 1957. Total mineral material, acidity, sulfur, and nitrogen in rain and snow at Kentville, Nova Scotia. *Tellus*, **9**:180–183.

Hill, A.C. 1971. Vegetation: a sink for atmospheric pollutants. *J. Air Poll. Contr. Assoc.*, **21**:341–346.

Hobbie, J.E. and G.E. Likens. 1973. The output of phosphorus dissolved organic carbon and fine particulate carbon from Hubbarb Brook watersheds. *Limmol. Oceanogr.*, **18**(5):734–742.

Hoeft, R.G., D.R. Keeney, and L.M. Walsh. 1972. Nitrogen and sulfur in precipitation and sulfur dioxide in the atmosphere in Wisconsin. *J. Environ. Qual.*, **1**:203–208

Holmes, R.T., and F.W. Sturges. 1973. Annual energy expenditure by the avifauna of a northern hardwoods ecosystem. *Oikos*, **24**:24–29.

Hooper, R.P., and C. Shoemaker. 1985. aluminum mobilization in an acidic headwater stream: temporal variation and mineral dissolution disequilibria. *Sciene*, **229**:464–465.

Hooper, R.S. 1986. The chemical response of an acid-sensitive headwater stream to snowmelt and storm events: a field study and simulation model. Ph.D. Thesis. Cornell University, Ithaca, N.Y. 279 pp.

Hornbeck, J.W., and G.E. Likens. 1974. The ecosystem concept for determining the importance of chemical composition of snow. pp. 139–151. In: *Advanced Concepts and Techniques in the Study of Snow and Ice Resources*. National Academy of Sciences, Washington, D.C.; Also, 31*st Annual Eastern Snow Conf. Proc.*, pp. 145–155.

Hornbeck, J.W., R.S. Pierce, and C.A. Federer. 1970. Streamflow changes after forest clearing in New England. *Water Resources Res.*, **6**(4):1124–1132.

Hornbeck, J.W., G.E. Likens, R.S. Pierce, and F.H. Bormann. 1975a. Strip cutting as a means of protecting site and streamflow quality when clearcutting northern hardwoods. pp. 209–229. In: B. Bernier and C.H. Winget (eds.), Proceedings of the 4th North American Forest Soils Conference on Forest Soils and Forest Land Management, August 1973, Quebec, Cananda. *Forest Soils and Forest Land Management*, University of Laval Press.

Hornbeck, J.W., R.S. Pierce, G.E. Likens, and C.W. Martin. 1975b. Moderating the impact of contemporary forest cutting on hydrologic and nutrient cycles. pp. 423–433. In: *International Symposium on the Hydrological Characteristics of Rive Basins*. Tokyo, December 1975, International Association of Hydrological Sciences Publ. 117.

Hornbeck, J.W., G.E. Likens, and J.S. Eaton. 1976. Seasonal variation in acidity of precipitation and the implications for forest-stream ecosystems. pp. 597–609. In: *Proceedings of the first International Symposium on Acid Precipitation and the Forest Ecosystem*. L.S. Dochinger and T.A. Seliga (eds.) U.S.D.A. For. Serv. Gen. Tech. Rep. Ne-23.

Hunt, C.B. 1967. Physiography of the United States. Freeman and Co., San Francisco, CA.

Huntington, T.G., C.E. Johnson, A.H. Johnson, T.G. Siccama, and D.F. Ryan. 1989. Carbon, organic matter and bulk density relationships in a forested Spodosol. *Soil Science*, **148**(5):380–386.

Hutchinson, G.E. 1957. *A Treatise on Limnology*. Vol. 1. J. Wiley and Sons, Inc. New York, NY. 1015 pp.

Johannessen, M., T. Dale, E.T. Gjessing, A. Henriksen, and R.F. Wright. 1976. *Proceedings of the International Symposium on Isotopes and Impurities in Snow and Ice*. International Association of Hydrological Science, Grenoble, France, August 28–30, 1975. International Association of Hydrological Science Publ. 118.

Johnson, C.E. 1989. The chemical and physical properties of a northern hardwood forest soil: harvesting effects, soil-tree relations and sample size determination. Ph.D. Thesis, University of Pennsylvania, Philadelphia, PA. 221 pp.

Johnson, N.M. 1971. Mineral equilibria in ecosystem geochemistry. *Ecology*, **52**:529–531.

Johnson, N.M., G.E. Likens, F.H. Bormann, and R.S. Pierce. 1968. Rate of chemical weathering of silicate minerals in New Hampshire. *Geochim. Cosmochim. Acta*, **32**:531–545.

Johnson, N.M., G.E. Likens, F.H. Bormann, D.W. Fisher, and R.S. Pierce. 1969. A working model for the variation in streamwater chemistry at the Hubbard Brook Experimental Forest, New Hampshire. *Water Resources Res.*, **5**(6):1353–1363.

Johnson, N.M., R.C. Reynolds, and G.E. Likens. 1972. Atmospheric sulfur: its effect on the chemical weathering of New England. *Science*, **177**:514–516.

Johnson, N.M., C.T. Driscoll, J.S. Eaton, G.E. Likens, and W.H. McDowell. 1981. "Acid rain," dissolved aluminum and chemical weathering at the Hubbard Brook Experimental Forest, New Hampshire. *Geochim. Cosmochim. Acta*, **45**(9):1421–1437.

Johnson, P.L., and W.T. Swank. 1973. Studies of cation budgets in the southern Appalachians on four experimental watersheds with contrasting vegetation. *Ecology*, **54**(1):70–80.

Jordan, C.F., and J.R. Kline. 1972. Relative stability of mineral cycles in forest ecosystems. *Amer. Nat.*, **106**(948):237–253.

Jordan, M.J., and G.E. Likens. 1975. An organic carbon budget for an oligotrophic lake in New Hampshire, U.S.A. *Verh. Intl. Verein. Limnol.*, **19**(2):994–1003.

Juang, F.H.T., and N.M. Johnson. 1967. Cycling of chlorine through a forested catchment in New England. *J. Geophys. Res.*, **72**:5641–5647.

Junge, C.E. 1958. The distribution of ammonia and nitrate in rain water over the United States. *Trans. Amer. Geophys. Union*, **39**:241–248.

Junge, C.E. 1963. *Air Chemistry and Radioactivity*. Academic Press, New York, NY. 382 pp.

Junge, C.E., and R.T. Werby. 1958. The concentration of chloride, sodium, potassium, calcium and sulfate in rain water over the United States. *J. Meteorol.*, **15**:417–425.

Küchler, A.W. 1964. Potential natural vegetation of the conterminous United States. American Geographical Society Spec. Publ. No. 36. New York, NY. 116 pp.

Lawrence, G.B., and C.T. Driscoll. 1990. Longitudinal patterns of concentration-discharge relationships in streamwater draining the Hubbard Brook Experimental Forest, New Hampshire. *J. Hydrology*, **116**:147–165.

Leonard, R.E. 1961. Interception of Precipitation by Northern Hardwoods. Station Paper No. 159, Northeastern Forest Experiment Station, U.S.D.A. Forest Service, Upper Darby, PA. 16 pp.

Leopold, A. 1949. *A Sand County Almanac*. Oxford University Press, Oxford, England. 226 pp.

Likens, G.E. 1972. *The Chemistry of Precipitation in the Central Finger Lakes Region*. Tech. Publ. No. 50, Cornell University Water Resources and Marine Sciences Center, Ithaca, New York. 47 pp. and 14 Figs.

Likens, G.E. 1973. *A Checklist of Organisms for the Hubbard Brook Ecosystems*. Section of Ecology and Systematics, Cornell University, Ithaca, NY. Mimeo. 54 pp.

Likens, G.E. 1989. Some aspects of air pollution effects on terrestrial ecosystems and prospects for the future. *Ambio*, **18**(3):172–178.

Likens, G.E. 1992. The Ecosystem Approach: Its Use and Abuse. *Excellence in Ecology*, Book 3. The Ecology Institute, Oldendorf-Luhe, Germany. 167 pp.

Likens, G.E., and F.H. Bormann. 1970. *Chemical Analyses of Plant Tissues from the Hubbard Brook Ecosystem in New Hampshire*. Bull. 79, Yale University School of Forestry, New Haven, CT. 25 pp.

Likens, G.E., and F.H. Bormann. 1972. Nutrient cycling in ecosystems. pp. 25–67. In: J. Wiens (ed.), *Ecosystems Structure and Function*. Oregon State University Press, Corvallis, OR.

Likens, G.E., and F.H. Bormann. 1974a. Linkages between terrestrial and aquatic ecosystems. *BioScience*, **24**(8):447–456.

Likens, G.E., and F.H. Bormann. 1974b. Acid rain: a serious regional environmental problem. *Science*, **184**(4143):1176–1179.

Likens, G.E., and F.H. Bormann. 1975. An experimental approach to nutrient-hydrologic interactions in New England landscapes. pp. 7–29. In: A.D. Hasler (ed.), *Proceedings of the INTECOL Symposium on Coupling of Land and Water Systems, 1971*. Leningrad. Springer-Verlag, New York, NY.

Likens, G.E., and M.B. Davis. 1975. Post-glacial history of Mirror Lake and its watershed in New Hampshire, U.S.A.: an initial report. *Verh. Intl. Verein. Limnol.*, **19**:982–993.

Likens, G.E., F.H. Bormann, N.M. Johnson, and R.S. Pierce. 1967. The calcium, magnesium, potassium and sodium budgets for a small forested ecosystem. *Ecology*, **48**:772–785.

Likens, G.E., F.H. Bormann, N.M. Johnson, D.W. Fisher, and R.S. Pierce. 1970. The effect of forest cutting and herbicide treatment on nutrient budgets in the Hubbard Brook watershed-ecosystem. *Ecol. Monogr.*, **40**(1):23–47.

Likens, G.E., F.H. Bormann, R.S. Pierce, and D.W. Fisher. 1971. Nutrient-hydrologic cycle interaction in small forested watershed-ecosystems. pp. 553–

563. In: P. Duvigneaud (ed.), *Productivity of Forest Ecosystems*, Proceedings of the Brussels Symposium, 1969. UNESCO, Paris, France.

Likens, G.E., F.H. Bormann, and N.M. Johnson. 1972. Acid rain. *Environment*, **14**(2):33–40.

Likens, G.E., N.M. Johnson, J.N. Galloway, and F.H. Bormann. 1976a. Acid precipitation: strong and weak acids. *Science*, **194**(4265):643–645.

Likens, G.E., F.H. Bormann, J.S. Eaton, R.S. Pierce, and N.M. Johnson. 1976b. Hydrogen ion input to the Hubbard Brook Experimental Forest, New Hampshire during the last decade. pp. 397–407. In: *Proceedings of the First International Symposium on Acid Precipitation and the Forest Ecosystem*, L.S. Dochinger and T.A. Seliga (eds.). U.S.D.A. For. Serv. Gen. Tech. Rep. NE-23.

Likens, G.E., E.S. Edgerton, and J.N. Galloway. 1983. The composition and deposition of organic carbon in precipitation. *Tellus*, **35B**:16–24.

Likens, G.E., F.H. Bormann, L.O. Hedin, C.T. Driscoll, and J.S. Eaton. 1990. Dry deposition of sulfur: a 23-yr record for the Hubbard Brook Forest Ecosystem. *Tellus*, **42B**:319–329.

Likens, G.E., C.T. Driscoll, D.C. Buso, T.G. Siccama, C.E. Johnson, G.M. Lovett, D.F. Ryan, T. Fahey, and W.A. Reiners. 1994. The biogeochemistry of potassium at Hubbard Brook. *Biogeochemistry*, **25**:61–125.

Likens, P.C. 1994. Publications of the Hubbard Brook Ecosystem Study. Institute of Ecosystem Studies, Millbrook, NY. 104 pp.

Lovett, G.M., G.E. Likens and S.S. Nolan. 1992. Dry deposition of sulfur to the Hubbard Brook Experimental Forest: A preliminary comparison of methods. pp. 1391–1402. In: *Precipitation Scavenging and Atmosphere-Surface Exchange*, Vol. 3, S.E. Schwartz and W.G.N. Slinn (eds.). Hemisphere Publ. Washington, DC.

Lyons, J.B., W.A. Bothner, R.H. Moench, and J.B. Thompson (eds.). 1995. Geologic map of New Hampshire. In Press.

Lunt, H.A. 1932. Profile characteristics of New England forest soils. *Conn. Agri. Exp. Sta. Bull.*, **342**:743–836.

MacIntyre, F. 1974. The top millimeter of the ocean. *Sci. Amer.*, **230**:67–77.

Marks, P.L. 1974. The role of pin cherry (*Prunus pensylvanica* L.) in the maintenance of stability in northern hardwood ecosystems. *Ecol. Monogr.*, **44**(1): 73–88.

Marshall, C.E. 1964. *The physical chemistry and mineralogy of soils*. Vol. 1. *Soil Materials*. Wiley, New York, NY. 388 pp.

Mast, M.A., and J.I. Drever. 1990. Chemical weathering in the Loch Vale watershed, Rocky Mountain National Park, Colorado. *Water Resour. Res.*, **26**:2971–2928.

Mau, D.P. 1993. Estimating ground water recharge and baseflow from streamflow hydrographs for a small Appalachian Mountain basin. M.S. Thesis, University of Colorado, Denver, CO. 130 pp.

McGarity, J.W., and J.A. Rajaratnam. 1973. Apparatus for the measurement of losses of nitrogen as gas from the field and simulated field environments. *Soil. Biol. Biochem.*, **5**:121–131.

Melillo, J.M. 1977. Nitrogen dynamics in an aggrading northern hardwood forest ecosystem. Ph.D. Thesis, Yale University, New Haven, CT.

Miller, R.B. 1963. Plant nutrients in hard beech—the immobilization of nutrients. *New Zealand J. Sci.*, **6**:365–377.

Moke, C.B. 1946. The geology of the Plymouth quadrangle, New Hampshire. New Hampshire Planning and Development Commission. Concord, NH. 21 pp.

Muller, R.N., and F.H. Bormann. 1976. The role of *Erythronium americanium* Ker. in energy flow and nutrient dynamics of a northern hardwood forest ecosystem. *Science*, **193**:1126–1128.

National Academy of Sciences. 1974. *United States Participation in the International Biological Program*. Rept. No. 6. Washington, DC. 166 pp.

Nodvin, S.C., C.T. Driscoll, and G.E. Likens. 1988. Soil processes and sulfate loss at the Hubbard Brook Experimental Forest. *Biogeochemistry*, **5**:185–199.

Northeastern Forest Experiment Station, U.S. Forest Service. 1972. Hubbard Brook Experimental Forest. Northeastern Forest Experiment Station, Upper Darby, PA. 13 pp.

Odum, E.P. 1959. *Fundamentals of Ecology*. 2nd Ed. W.B. Saunders Co., Philadelphia, PA. 546 pp.

Oosting, H.J. 1956. *The Study of Plant Communities*. W.H. Freeman Co., San Francisco, CA. 440 pp.

Paillet, F.L., A.E. Hess, C.H. Cheng, and E. Hardin. 1987. Characterization of fracture permeability with high-resolution vertical flow measurements during bolehole pumping. *J. Ground Water*, **25**(1):28–40.

Pearson, F.J., and D.W. Fisher. 1971. Chemical Composition of Atmospheric Precipitation in the Northeastern United States. United States Geological Survey Water Supply Paper 1535-P. Washington, DC. pp. P1–P23.

Penman, H.L. 1956. Estimating evapotranspiration. *Trans. Amer. Geophys. Union*, **37**:43–50.

Pierce, R.S. 1967. Evidence of overland flow on forest watersheds. pp. 247–252. In: W.E. Sopper and H.W. Lull (eds.), *International Symposium on Forest Hydrology*. Pergamon Press, New York, NY.

Pierce, R.S., J.W. Hornbeck, G.E. Likens, and F.H. Bormann. 1970. Effects of elimination of vegetation on stream water quantity and quality. pp. 311–328. In: Proc. International Association Scientific Hydrology. UNESCO. *Results of Research on Representative and Experimental Basins*, Wellington, New Zealand, December 1970.

Pierce, R.S., C.W. Martin, C.C. Reeves, G.E. Likens, and F.H. Bormann. 1972. Nutrient loss from clearcuttings in New Hampshire. *Symposium on Watersheds in Trans.*, Ft. Collins, Colorado. pp. 285–295. American Water Resources Assoc. and Colorado State Univ.

Polynov, B.B. 1937. *The cycle of weathering*. (A. Muir, Transl.) Thomas Murby. London, England.

Powell, S.T. 1964. Quality of water. pp. 19-1 to 19-37. In: V.T. Chow (ed.), *Handbook of Applied Hydrology*. McGraw-Hill Book Co., New York, NY.

Reuss, J.O. 1976. Chemical and biological relationships relevant to the ecological effects of acid rainfall. pp. 791–813. In: *Proceedings of the First International Symposium on Acid Precipitation and the Forest Ecosystem*. L.S. Dochinger and T.A. Seliga (eds.). U.S.D.A. For. Serv. Gen. Tech. Rep. NE-23.

Reynolds, R.C., and N.M. Johnson. 1972. Chemical weathering in the temperate glacial environment of the northern Cascade Mountains. *Geochim. Cosmochim. Acta*, **36**:537–554.

Roskoski, J.P. 1975. Differential nitrogen fixation in wood litter. *Bull. Ecol. Soc. Amer.*, **56**(2):12.

Schindler, D.W., and J.E. Nighswander. 1970. Nutrient supply and primary production in Clear Lake, Eastern Ontario. *J. Fish. Res. Bd. Can.*, **27**(11): 2009–2036.

Schindler, D.W., R.W. Newbury, K.G. Beaty, and P. Campbell. 1976. Natural water and chemical budgets for a small Precambrian lake basin in central Canada. *J. Fish. Res. Bd. Canada*, **33**:2526–2543.

Schofield, C.L. 1976. Lake acidification in the Adirondack Mountains of New York: causes and consequences. p. 477. In: *Proceedings of the First International Symposium on Acid Precipitation and the Forest Ecosystem.* L.S. Dochinger and T.A. Seliga (eds.). U.S.D.A. For. Serv. Gen. Tech. Rep. NE-23.

Scrivener, J.C. 1975. Water, water chemistry and hydrochemical balance of dissolved ions in Carnation Creek watershed, Vancouver Island, July 1971–May 1974. Technical Report 564, Fisheries Marine Services Research and Development. Vancouver, BC. 141 pp.

Sellers, W.D. 1965. *Physical Climatology*. The University of Chicago Press, Chicago, IL. 272 pp.

Shattuck, P.C. 1991. Shallow water-table response to precipitation and evapotranspiration in an ephemeral stream valley, Woodstock, New Hampshire. M.S. Thesis, University of New Hampshire, Durham, NH. 113 pp.

Siccama, T.G., F.H. Bormann, and G.E. Likens. 1970. The Hubbard Brook Ecosystem Study: productivity, nutrients and phytosociology of the herbaceous layer. *Ecol. Monogr.* **40**(4):389–402.

Siegler, H.R. (ed.). 1968. *The White-tailed Deer of New Hampshire*. New Hampshire Fish and Game Dept., Concord, NH. 256 pp.

Smith, W.H. 1976. Character and significance of forest tree root exudates. *Ecology*, **57**:324–331.

Sopper, W.E., and H.W. Lull, 1965. The representativeness of small forested experimental watersheds in northeastern United States. *Intl. Assoc. Sci. Hydrol.*, **66**(2):441–456.

Sopper, W.E., and H.W. Lull. 1970. *Streamflow Characteristics of the Northeastern United States*. Bull. 766. The Pennsylvania State University, Univ. Park, PA. 129 pp.

Streets, D.G., and T.D. Veselka. 1987. Future emissions. In: *Interim Assessment of the National Acid Precipitation Assessment Program*, pp. 3-1 to 3-33. Washington, DC.

Stresky, S.J. 1991. Morphology and flow characteristics of pipes in a forested New England hillslope. M.S. Thesis, University of New Hampshire, Durham, NH. 131 pp.

Stumm, W., and J.J. Morgan. 1970. *Aquatic Chemistry*. Wiley-Interscience, New York, NY. 583 pp.

Sturges, F.W., R.T. Holmes, and G.E. Likens. 1974. The role of birds in nutrient cycling in a northern hardwoods ecosystem. *Ecology*, **55**(1):149–155.

Tamm, C.O., and T. Troedsson. 1955. An example of the amounts of plant nutrients supplied to the ground in road dust. *Oikos*, **6**:61–70.

Taylor, A.W., W.M. Edwards, and E.C. Simpson. 1971. Nutrients in streams draining woodland and farm land near Coshocton, Ohio. *Water Resources Res.*, **7**(1):81–89.

Thiessen, A.H. 1911. Precipitation for large areas. *Monthly Weather Rev.*, **39**:1082–1089.

Thornthwaite, C.W. 1948. An approach toward a rational classification of climate. *Geogr. Rev.*, **38**:55–94.

Trewartha, G.T. 1954. *Introduction to climate.* McGraw-Hill Book Co., New York, NY. 402 pp.

Ungemach, H. 1967. pp. 221–226. In: H. Lent (ed.), *Atas do Simpósio sâbre a Biota Amazônica.* Vol. 3. Rio de Janeiro, Brazil.

Ungemach, H. 1970. Chemical rain water studies in the Amazon region. pp. 354–358. In: J.M. Idrobo (ed.), *Simpósio Y Ford de Biologia Tropical Amazônica.* Vol. II. Editorial PAX, Bogotá, Colombia.

Viro, P.J. 1953. Loss of nutrients and the natural nutrient balance of the soil in Finland. *Comm. Inst. Forest Fenn.*, **42**:1–50.

White, E.J., and F. Turner. 1970. A method of estimating income of nutrients in catch of airborne particles by a woodland canopy. *J. Appl. Ecol.*, **7**:441–461.

Whittaker, R.H., and G.E. Likens. 1973. Carbon in the biota. pp. 281–302. In: G.M. Woodwell and E.V. Pecan (eds.), *Carbon and the Biosphere.* CONF-720510, United States Atomic Energy Commission, Springfield, VA.

Whittaker, R.H., and G.E. Likens. 1975. The biosphere and man. pp. 305–328. In: H. Lieth and R.H. Whittaker (eds.), *Primary Productivity of the Biosphere.* Springer-Verlag, NY.

Whittaker, R.H., F.H. Bormann, G.E. Likens, and T.G. Siccama. 1974. The Hubbard Brook Ecosystem Study: forest biomass and production. *Ecol. Monogr.*, **44**(2):233–254.

Williams, A.F., L. Ternan, and M. Kent. 1986. Some observations on the chemical weathering of the Dartmoor granite. *Earth Surface Processes and Landforms*, **11**:557–574.

Winter, T.C. 1985. Approaches to the study of lake hydrology. pp. 128–135. In: An Ecosystem Approach to Aquatic Ecology: Mirror Lake and its Environment, G.E. Likens (ed.). Springer-Verlag, New York, NY.

Winter, T.C., J.S. Eaton, and G.E. Likens. 1989. Evaluation of inflow to Mirror Lake, New Hampshire. *Water Resour. Bull.*, **25**(5):991–1008.

Wood, T., and F.H. Bormann. 1974. The effects of an artificial acid mist upon the growth of *Betula alleghaniensis* Britt. *Environ. Pol.*, **7**:259–268.

Wood, T., and F.H. Bormann. 1975. Increases in foliar leaching caused by acidification of an artificial mist. *Ambio*, **4**(4):169–171.

Woodwell, G.M., and R.H. Whittaker. 1967. Primary production and the cation budget of the Brookhaven Forest. pp. 151–166. In: H.E. Young (ed.), *Symposium on Primary Productivity and Mineral Cycling in Natural Ecosystems.* University of Maine Press, Orono, ME.

Wright, R.F. 1974. *Forest Fire: Impact on the Hydrology, Chemistry and Sediments of Small Lakes in Northeastern Minnesota.* Interim Rept. No. 10, Limnology Research Center, University of Minnesota, Minneapolis, MN. 129 pp.

Wright, R.F., T. Dale, E.T. Gjessing, G.R. Hendrey, A. Henriksen, M. Johannessen, and I.P. Muniz. 1976. Impact of acid precipitation on fresh-water ecosystems in Norway. pp. 459–476. In: *Proceedings of the First International Symposium on Acid Precipitation and the Forest Ecosystem*, L.S. Dochinger and T.A. Seliga (eds.). U.S.D.A. For. Serv. Gen. Tech. Rep. NE-23.

Yuretich, R., and G. Batchelder. 1988. Hydrogeochemical cycling and chemical denudation in the Fort River watershed, central Massachusetts: an appraisal of mass balance studies. *Water Resour. Res.*, **24**:105–114.

Selected References for Further Reading About the Hubbard Brook Ecosystem Study

Hydrology

Lawrence, G.B., C.T. Driscoll, and R.D. Fuller. (1988). Hydrologic control of aluminum chemistry in an acidic headwater stream. *Water Resour. Res.*, **24**(5): 659–669.

Federer, C.A., L.D. Flynn, C.W. Martin, J.W. Hornbeck, and R.S. Pierce. (1990). Thirty years of hydrometeorologic data at the Hubbard Brook Experimental Forest, New Hampshire. USDA Forest Service, General Technical Report NE-141. 44 pp.

Atmospheric Deposition

Acid Rain

Johnson, N.M., C.T. Driscoll, J.S. Eaton, G.E. Likens, and W.H. McDowell. (1981). "Acid rain," dissolved aluminum and chemical weathering at the Hubbard Brook Experimental Forest, New Hampshire. *Geochim. Cosmochim. Acta*, **45**(9):1421–1437.

Bormann, F.H. (1982). The effects of air pollution on the New England landscape, Part I, *Ambio* **11**(4):188–194; Part II, *Ambio*, **11**(6):338–346.

Munn, R.E., G.E. Likens, B. Weisman, J.W. Hornbeck, C.W. Martin, and F.H. Bormann. (1984). A meteorological analysis of the precipitation chemistry event samples at Hubbard Brook, New Hampshire. *Atmos. Environ.*, **18**(12): 2775–2779.

Driscoll, C.T., N.M. Johnson, G.E. Likens, and M.C. Feller. (1988). Effects of acidic deposition on the chemistry of headwater streams: a comparison between Hubbard Brook, New Hampshire, and Jamieson Creek, British Columbia. *Water Resour. Res.*, **24**(2):195–200.

Weathers, K.C., G.E. Likens, F.H. Bormann, S.H. Bicknell, B.T. Bormann, B.C. Daube, Jr., J.S. Eaton, J.N. Galloway, W.C. Keene, K.D. Kimball, W.H. McDowell, T.G. Siccama, D. Smiley, and R.A. Tarrant. (1988). Cloudwater chemistry from ten sites in North America. *Environ. Sci. Technol.*, **22**(8):1018–1026.

Likens, G.E. (1989). Some aspects of air pollution effects on terrestrial ecosystems and prospects for the future. *Ambio*, **18**(3):172–178.

Hornbeck, J.W. (1992). Comparative impacts of forest harvest and acid precipitation on soil and streamwater acidity. *Environ. Pollution*, **7**(2/3):151–155.

Dry Deposition

Eaton, J.S., G.E. Likens, and F.H. Bormann. (1978). The input of gaseous and particulate sulfur to a forest ecosystem. *Tellus*, **30**:546–551.

Likens, G.E., F.H. Bormann, L.O. Hedin, C.T. Driscoll, and J.S. Eaton. (1990). Dry deposition of sulfur: a 23-yr record for the Hubbard Brook Forest Ecosystem. *Tellus*, **42B**:319–329.

Lovett, G.M., G.E. Likens, and S.S. Nolan. (1992). Dry deposition of sulfur to the Hubbard Brook Experimental Forest: A preliminary comparison of methods. In: *Fifth Internat. Conf. on Precipitation Scavenging and Atmosphere-Surface Exchange*. S.E. Schwartz and W.G.N. Slinn (coordinators). pp. 1391–1402. *The Summers Volume (3): Applications and Appraisals. Hemisphere Publ.*, Washington, DC.

Gaseous Emissions

Bowden, W.B. (1986). Gaseous nitrogen emissions from undisturbed terrestrial ecosystems: an assessment of their impacts on local and global nitrogen budgets. *Biogeochemistry*, **2**(3):249–279.

Bowden, W.B., and F.H. Bormann. (1986). Transport and loss of nitrous oxide in soil water after forest clear cutting. *Science*, **233**:867–869.

Soil Chemistry

Driscoll, C.T., N. van Breemen, and J. Mulder. (1985). Aluminum chemistry in a forested spodosol. *Soil Sci. Soc. Amer. J.*, **49**(2):437–444.

Fuller, R.D., M.B. David, and C.T. Driscoll. (1985). Sulfate adsorption relationships in forested spodosols of the northeastern USA. *Soil Sci. Soc. Amer. J.*, **49**(4):1034–1040.

Nodvin, S.C., C.T. Driscoll, and G.E. Likens. (1986). The effect of pH on sulfate adsorption by a forest soil. *Soil Science*, **142**(2):69–75.

Nodvin, S.C., C.T. Driscoll, and G.E. Likens. (1986). Simple partitioning of anions and dissolved organic carbon in a forest soil. *Soil Science*, **142**(1):27–35.

Huntington, T.G., D.F. Ryan, and S.P. Hamburg. (1988). Estimating soil nitrogen and carbon pools in a northern hardwood forest ecosystem. *Soil Sci. Soc. Am. J.*, **52**:1162–1167.

Huntington, T.G., C.E. Johnson, A.H. Johnson, T.G. Siccama, and D.F. Ryan. (1989). Carbon, organic matter and bulk density relationships in a forested Spodosol. *Soil Science*, **148**(5):380–386.

Weathering

Bailey, S.W. (1994). Biogeochemistry of aluminum and calcium in a linked forest-aquatic ecosystem. Ph.D. Thesis, Syracuse University, Syracuse, NY. 86 pp.

Likens, G.E., C.T. Driscoll, D.C. Buso, T.G. Siccama, C.E. Johnson, D.F. Ryan, G.M. Lovett, T. Fahey, and W.A. Reiners. (1994). The biogeochemistry of potassium at Hubbard Brook. *Biogeochemistry*, **25**:61–125.

Chemical Flux and Cycling

General

Muller, R.N. (1978). The phenology, growth and ecosystem dynamics of *Erythronium americanum* in the northern hardwood forest. *Ecol. Monogr.*, **48**(1): 1–20.

Whittaker, R.H., G.E. Likens, F.H. Bormann, J.S. Eaton, and T.G. Siccama. (1979). The Hubbard Brook Ecosystem Study: forest nutrient cycling and element behavior. *Ecology*, **60**(1):203–220.

Ryan, D.F., and F.H. Bormann. (1982). Nutrient resorption in northern hardwood forests. *BioScience*, **32**(1):29–32.

McDowell, W.H., and G.E. Likens. (1988). Origin, composition, and flux of dissolved organic carbon in the Hubbard Brook Valley. *Ecol. Monogr.*, **58**(3): 177–195.

Lawrence, G.B., and C.T. Driscoll. (1990). Longitudinal patterns of concentration-discharge relationships in streamwater draining the Hubbard Brook Experimental Forest, New Hampshire. *J. Hydrology*, **116**:147–165.

Nitrogen

Bormann, F.H., G.E. Likens, and J.M. Melillo. (1977). Nitrogen budget for an aggrading northern hardwood forest ecosystem. *Science*, **196**(4293):981–983.

Roskoski, J.P. (1980). Nitrogen fixation in a northern hardwood forest in the northeastern United States. *Plant and Soil*, **54**:33–44.

Melillo, J.M., J.D. Aber, and J.F. Muratore. (1982). Nitrogen and lignin control of hardwood leaf litter decomposition dynamics. *Ecology*, **63**(3):621–626.

Federer, C.A. (1983). Nitrogen mineralization and nitrification: depth variation in four New England forest soils. *Soil Sci. Soc. Amer. J.*, **47**:1008–1014.

Duggin, J.A., G.K. Voigt, and F.H. Bormann. (1991). Autotrophic and heterotrophic nitrification in response to clear-cutting northern hardwood forest. *Soil Biol. Biochem.*, **23**(8):779–787.

Bormann, B.T., F.H. Bormann, W.B. Bowden, R.S. Pierce, S.P. Hamburg, D. Wang, M.C. Snyder, C.Y. Li, and R.C. Ingersoll. (1993). Rapid N_2 fixation in pines, alder and locust: evidence from the sandbox ecosystem study. *Ecology*, **74**(2):583–598.

Phosphorus

Wood, T., F.H. Bormann, and G.K. Voigt. (1984). Phosphorus cycling in a northern hardwood forest: biological and chemical control. *Science*, **223**: 391–393.

Yanai, R.D. (1992). Phosphorus budget of a 70-yr-old northern hardwood forest. *Biogeochemistry*, **17**:1–22.

Sulfur

Fuller, R.D., M.B. David, and C.T. Driscoll. (1985). Sulfate adsorption relationships in forested spodosols of the northeastern USA. *Soil Sci. Soc. Amer. J.*, **49**(4):1034–1040.

Schindler, S.C., M.J. Mitchell, T.J. Scott, R.D. Fuller, and C.T. Driscoll. (1986). Incorporation of ^{35}S-sulfate into inorganic and organic constituents of two forest soils. *Soil Sci. Soc. Am. J.*, **50**:457–462.

Nodvin, S.C., C.T. Driscoll, and G.E. Likens. (1988). Soil processes and sulfate loss at the Hubbard Brook Experimental Forest. *Biogeochemistry*, **5**:185–199.

Mitchell, M.J., M.B. David, and R.B. Harrison. (1992). Sulphur dynamics of forest ecosystems. pp. 215–254. In: *Sulphur Cycling on the Continents*. R.W. Howarth, J.W.B. Stewart and M.V. Ivanov (eds). SCOPE Vol. 48. John Wiley & Sons, New York, NY.

Aluminum

Driscoll, C.T., N. van Breemen, and J. Mulder. (1985). Aluminum chemistry in a forested spodosol. *Soil Sci. Soc. Amer. J.*, **49**(2):437–444.

Hooper, R.P., and C. Shoemaker. (1985). Aluminum mobilization in an acidic headwater stream: temporal variation and mineral dissolution disequilibria. *Science*, **229**:463–465.

Lawrence, G.B., R.D. Fuller, and C.T. Driscoll. (1986). Spatial relationships of aluminum chemistry in the streams of the Hubbard Brook Experimental Forest, New Hampshire. *Biogeochemistry*, **2**(2):115–135.

Driscoll, C.T., and D.A. Schaefer. (1990). Background on Nitrogen Processes. In: *The Role of Nitrogen in the Acidification of Soils and Surface Waters*. J.L. Malanchuk and J. Nilsson (eds). pp. 4.1 to 4.2. Nordic Council of Ministers, Solna, Sweden.

Browne, B.A., and C.T. Driscoll. (1992). Soluble aluminum silicates: stoichiometry, stability and implications for environmental geochemistry. *Science*, **256**: 1667–1670.

Cations

Driscoll, C.T., and G.E. Likens. (1982). Hydrogen ion budget of an aggrading forested ecosystem. *Tellus*, **34**:283–292

Likens, G.E., C.T. Driscoll, D.C. Buso, T.G. Siccama, C.E. Johnson, D.F. Ryan, G.M. Lovett, T. Fahey, and W.A. Reiners. (1994). The biogeochemistry of potassium at Hubbard Brook. *Biogeochemistry*, **25**:61–125.

Animals

Strayer, D., D.H. Pletscher, S.P. Hamburg, and S.C. Nodvin. (1986). The effects of forest disturbance on land gastropod communities in northern New England. *Can. J. Zool.*, **64**:2094–2098.

Pletscher, D.H. (1987). Nutrient budgets for white-tailed deer in New England with special reference to sodium. *J. Mamm.*, **68**(2):330–336.

Pletscher, D.H., F.H. Bormann, and R.S. Miller. (1989). Importance of deer compared to other vertebrates in nutrient cycling and energy flow in a northern hardwood ecosystem. *Amer. Midl. Natur.*, **121**:302–311.

Long-Term Trends

Likens, G.E., F.H. Bormann, R.S. Pierce, J.S. Eaton, and R.E. Munn. (1984). Long-term trends in precipitation chemistry at Hubbard Brook, New Hampshire. *Atmos. Environ.*, **18**(12):2641–2647.

Driscoll, C.T., G.E. Likens, L.O. Hedin, J.S. Eaton, and F.H. Bormann. (1989). Changes in the chemistry of surface waters: 25-year results at the Hubbard Brook Experimental Forest, NH. *Environ. Sci. Technol.*, **23**(2):137–143.

Federer, C.A., J.W. Hornbeck, L.M. Tritton, C.W. Martin, R.S. Pierce, and C.T. Smith. (1989). Long-term depletion of calcium and other nutrients in eastern U.S. forests. *Environ. Management*, **13**(5):593–601.

Likens, G.E., L.O. Hedin, and T.J. Butler. (1990). Some long-term precipitation chemistry patterns at the Hubbard Brook Experimental Forest: extremes and averages. *Verh. Internat. Verein. Limnol.*, **24**(1):128–135.

Hedin, L.O., L. Granat, G.E. Likens, T.A. Buishand, J.N. Galloway, T.J. Butler, and H. Rodhe. (1994). Steep declines in atmospheric base cations in regions of Europe and North America. *Nature*, **367**:351–354.

Trace Metals

Siccama, T.G., and W.H. Smith. (1978). Lead accumulation in a northern hardwood forest. *Environ. Sci. Technol.*, **12**:593–594.

Smith, W.H., and T.G. Siccama. (1981). The Hubbard Brook Ecosystem Study: biogeochemistry of lead in the northern hardwood forest. *J. Environ. Qual.*, **10**(3):323–333.

Driscoll, C.T., R.D. Fuller, and D.M. Simone. (1988). Longitudinal variations in trace metal concentrations in a northern forested ecosystem. *J. Environ. Qual.*, **17**(1):101–107.

Johnson, C.E., T.G. Siccama, C.T. Driscoll, G.E. Likens, and R.E. Moeller. (1995). Changes in forest lead cycling in response to decreasing atmospheric inputs. Ecol. Appl. (In Press).

Stream Ecosystems

Meyer, J.L., and G.E. Likens. (1979). Transport and transformation of phosphorus in a forest stream ecosystem. *Ecology*, **60**(6):1255–1269.

Hall, R.J., G.E. Likens, S.B. Fiance, and G.R. Hendrey. (1980). Experimental acidification of a stream in the Hubbard Brook Experimental Forest, New Hampshire. *Ecology*, **61**(4):976–989.

Bilby, R.E. (1981). Role of organic debris dams in regulating the export of dissolved and particulate matter from a forested watershed. *Ecology*, **62**(5): 1234–1243.

Meyer, J.L., G.E. Likens, and J. Sloane. (1981). Phosphorus, nitrogen, and organic carbon flux in a headwater stream. *Arch. Hydrobiol.*, **91**(1):28–44.

Likens, G.E., and R.E. Bilby. (1982). Development, maintenance and role of organic-debris dams in New England streams. pp. 122–128. In: *Sediment Budgets and Routing in Forested Drainage Basins.* F.J. Swanson, R.J. Janda, T. Dunne and D.N. Swanston (eds.). U.S.D.A. Forest Service General Technical Report PNW-141.

Hall, R.J., C.T. Driscoll, and G.E. Likens. (1987). Importance of hydrogen ions and aluminium in regulating the structure and function of stream ecosystems: an experimental test. *Freshwater Biology*, **18**:17–43.

Mayer, M.S., and G.E. Likens. (1987). The importance of algae in a shaded headwater stream as food for an abundant caddisfly (Trichoptera). *J.N. Am. Benthol. Soc.*, **6**(4):262–269.

Hedin, L.O. (1990). Factors controlling sediment community respiration in woodland stream ecosystems. *Oikos*, **57**:94–105.

Simulation Modeling

Federer, C.A., and D. Lash. (1978). BROOK: A hydrologic simulation model for eastern forests. Water Resour. Res. Center, University of New Hampshire, Durham, NH. Research Report No. 19. 84 pp.

Fuller, R.D., C.T. Driscoll, S.C. Schindler, and M.J. Mitchell. (1986). A simulation model of sulfur transformations in forested Spodosols. *Biogeochemistry*, **2**(4):313–328.

Pu Mou, and T.J. Fahey. (1993). REGROW: a computer model simulating the early successional process of a disturbed northern hardwood ecosystem. *J. Appl. Ecology*, **30**:676–688.

Mirror Lake Ecosystem

Caraco, N., J.J. Cole, G.E. Likens, M.D. Mattson, and S. Nolan. (1988). A very imbalanced nutrient budget for Mirror Lake, New Hampshire, USA. *Verh. Internat. Verein. Limnol.*, **23**(1):170–175.

Cole, J.J., N.F. Caraco, and G.E. Likens. (1990). Short-range atmospheric transport: significant source of phosphorus to an oligotrophic lake. *Limnol. Oceanogr.*, **35**(6):1230–1237.

Giblin, A.E., G.E. Likens, D. White, and R.W. Howarth. (1990). Sulfur storage and alkalinity generation in New England lake sediments. *Limnol. Oceanogr.*, **35**(4):852–869.

Mattson, M.D., and G.E. Likens. (1990). Air pressure and methane fluxes. *Nature*, **347**(6295):718–719.

Caraco, N.F., J.J. Cole, and G.E. Likens. (1992). New and recycled primary production in an oligotrophic lake: Insights for summer phosphorus dynamics. *Limnol. Oceanogr.*, **37**(3):590–602.

Mattson, M.D., and G.E. Likens. (1993). Redox reactions of organic matter decompositon in a soft water lake. *Biogeochemistry*, **19**:149–172.

Other Hubbard Brook Books

In *Pattern and Processes in a Forested Ecosystem* (Bormann, F.H., and G.E. Likens, 1979), we explore how biogeochemical and other ecosystem processes respond to disturbance (forest cutting) and how these processes recover as the ecosystem redevelops and moves toward a predisturbance condition.

In *An Ecosystem Approach to Aquatic Ecology: Mirror Lake and its Environment* (Likens, G.E. ed, 1985) biogeochemical interactions among Mirror Lake, its airshed, its watershed, and the larger forested landscape of the Hubbard Brook Valley are evaluated, as are the ecology and the internal biogeochemical processes of the lake itself. These features are considered within an historical perspective dating from the origin of the lake some 14,000 yr B.P.

In the *Ecosystem Approach*: *Its Use and Abuse* (Likens, G.E. 1992), approaches to the study of ecosystem ecology are outlined and examples of sustained ecological research from the Hubbard Brook Ecosystem Study are given. Biogeochemical linkages between air, land, and water provide a context for assessing the interactions between ecosystem ecology and societal concerns, particularly as related to complex environmental issues.

Subject Index

References to the principal treatment of a subject (section headings) are set in **boldface.**

O

P

R